MY COMET TALES

NICK B. COMANDE

authorHOUSE

AuthorHouse™
1663 Liberty Drive
Bloomington, IN 47403
www.authorhouse.com
Phone: 833-262-8899

© 2024 Nick B. Comande. All rights reserved.

No part of this book may be reproduced, stored in a retrieval system, or transmitted by any means without the written permission of the author.

Published by AuthorHouse 10/01/2024

ISBN: 979-8-8230-3008-3 (sc)
ISBN: 979-8-8230-3009-0 (hc)
ISBN: 979-8-8230-3007-6 (e)

Library of Congress Control Number: 2024914397

Print information available on the last page.

Any people depicted in stock imagery provided by Getty Images are models, and such images are being used for illustrative purposes only.
Certain stock imagery © Getty Images.

This book is printed on acid-free paper.

Because of the dynamic nature of the Internet, any web addresses or links contained in this book may have changed since publication and may no longer be valid. The views expressed in this work are solely those of the author and do not necessarily reflect the views of the publisher, and the publisher hereby disclaims any responsibility for them.

To Rachel, the brightest star in my galaxy.

Contents

Introduction ix

Chapter 1	....	1
Chapter 2	Preparing for Halley's Comet	6
Chapter 3	November 20, 1985	13
Chapter 4	December 6, 1985	18
Chapter 5	January 11, 1986	20
Chapter 6	January 13, 1986	25
Chapter 7	January 14, 1986	28
Chapter 8	January and February 1986	31
Chapter 9	March 16, 1986	36
Chapter 10	March 17, 1986	40
Chapter 11	March 18, 1986	44
Chapter 12	March 19, 1986	47
Chapter 13	March 20, 1986	51
Chapter 14	March 21, 1986	55
Chapter 15	March 22, 1986	59
Chapter 16	March 28, 1986	63
Chapter 17	March 29, 1986	66
Chapter 18	April 6, 1986	69
Chapter 19	April 16, 1986	72
Chapter 20	April 22, 1986	75
Chapter 21	April 28, 1986	78
Chapter 22	In the Wake of the Comet	82
Chapter 23	Halley's Comet Facts	88

Introduction

There are a countless number of stars in the sky. There are also moons, planets, nebulas, galaxies, and star clusters. It would help if you had a good telescope to see many of these, but there are other things in the night sky that you don't need a telescope to see. The most obvious is the Earth's moon, our closest neighbor is roughly a quarter million miles away. Its constantly changing phases are silently responsible for influencing our tides, and when full, it is bright enough to cast shadows on the ground. There are stars of different brightness that seem to twinkle in the night sky and while you see them, you may not notice the larger stars are really the planets of our own solar system. What we call shooting stars are not stars at all, but small meteorites that leave a brief fiery trail as they skip off or burn up as they enter the Earth's atmosphere.

Then there are comets—remnants of the solar system's formation made up of frozen rock, dust, ice, and gases. There are thousands of them out there, but there is one that stands out when you hear the word comet, and that is Halley's Comet. What is Halley's Comet? First, its official name is 1P/Halley. The P stands for periodic, and the 1 stands for it being the first comet identified as a periodic. It is the most famous comet in history because it is a periodic comet, meaning it returns in a highly elliptical orbit around our sun every 74-79 years with the average being 76 years. The change in year's is due to the gravitation pull and push of the larger planets such as Jupiter which can slow it down or speed it up. This qualifies Halley to be called a short-period comet and makes it possible for some people to see it twice a lifetime, providing you are very young the first time you see it. It is a giant snowball made up of rock and ice. The ices on comets are composed mostly of water, carbon dioxide, and ammonia. This giant rock in space measures some 9.3 miles by 5 miles in size, orbits around our sun, and travels past Neptune's orbit before returning. When a comet, like Halley's, comes back from the deep reaches of space and approaches the sun, its surface begins to heat up, and the ice sublimates and reflects sunlight to create an immensely long tail. It grows in length as it nears the sun and can reach up to 50 million miles long when closest to the sun. It will bright enough to be seen with the naked eye if the area you are in is dark enough.

Oddly enough, this comet was named after someone who never saw it. We must thank the English astronomer Edmond Halley, who in 1705 calculated the past appearances of this particular comet dating back to 240 BC and accurately predicted its return in late 1758, earning the honor of having the comet named after him. Halley was also a brilliant mathematician, inventor and engineer. Unfortunately, Edmond (or Edmund) Halley died in January of 1742, 16 years before its subsequent return. Halley's Comet's most recent appearances were in 1910 and 1985-86, with its next viewing in 2061. I was fascinated by Halley's Comet during its 1985-86 return. Why I kept notes on each night I viewed it, I really don't remember why. Perhaps because it was a once in a lifetime event for me. I recently came across these notes and put them together to create these comet tales.

Chapter 1

So, where were you when you first heard of Halley's Comet? I can pinpoint the day. Being born in the late 50s and a child of the 60s, I grew up in a time when imaginations ran wild, and the world was a much different place to grow up in. Friends could play outside late at night and didn't have to worry about the problems they would eventually face as adults.

I grew up with an imagination that was fed with the world of science fiction, thanks to the television shows of Irwin Allen. Some of my favorite Allen's shows of the mid-60s included the sci-fi classics like *Lost in Space, Land of the Giants*, and *The Time Tunnel*. *Lost in Space* followed the Robinson family and their adventures in far-away galaxies, inspiring my fascination with space from a very young age. To me, it was always a pleasure to look up at the night sky, a black canvas that was covered with an infinite number of lights- to see the moon and the stars and, if lucky, occasionally see a shooting star streak across the sky. Some were bright, some not so bright, while others seemed to twinkle or move nightly. I would later learn that some of those stars were not stars but the other planets of our solar system. The different phases of the moon also piqued my interest, and I loved seeing it change and grow a little more every night until it was completely full, sometimes even casting the reflective light of the sun so that some nights were so bright you could see your shadow beside you on the ground. There was so much to see in the night sky, and all you had to do was look up!

Some of the most beautiful skies I have ever seen were when the moon was full or nearly full, with patches of clouds in the moon's foreground that showed various grey colors in the sky, backlit by the moon's light. As a child, on warm summer nights, I enjoyed sitting in my yard, just looking up at that particular type of moon and cloud-filled sky, imagining I was in a rocket on my way to the moon and beyond. It was summer nights like these that Ray Bradbury wrote about his novels back in the 1950s. These, of course, were the very same novels I would read in the 70s. His novels, and those of other sci-fi writers like H.G. Wells and Jules Verne, were only part of the things that made me think about the wonders of space. Yet, there was something else during the 60s that sparked my curiosity. The United States of America, with thanks to President John F. Kennedy, pushed forward the proposal to put a man on the moon and return him safely to Earth by the end of the decade. But there is still another time to talk about that as I have gotten a bit off-topic since I first asked, "Where were you when you first heard of Halley's Comet"? Thanks to the internet, I was able to find out.

The first time I heard about Halley's Comet was the night of September 23, 1966, between 7:00 and 7:30 pm CST, during an episode of Irwin Allen's show *The Time Tunnel*. It was the third episode of the series titled "End of the World." It told a story set in 1910, with people living in a small town who were afraid the world was coming to an end with the approach of Halley's Comet. While factually wrong, considering the world didn't end and we are still here, people were fearful of something they didn't fully understand. In 1910, there were some who thought Halley's Comet would possibly cause the end of the world. Similar confusion occurred when people thought Martians were invading the world during Orson Welles' famous *War of the Worlds* (1938) radio broadcast. I was very young when the Halley's Comet episode of *The Time Tunnel* aired and had no knowledge (at least none I can remember) of the comet. To me, *The Time Tunnel* was just an exciting show, and I let the notion of the comet go. As years passed and I went to school, as all children did, I eventually learned more about the comet and how it returned every 76 years.

Somewhere along the line, it dawned on me that even though the return of Halley's Comet was still years away, I realized I would be able to see it during one of its voyages through our solar system. I will confess I do not remember exactly when I became, and I hate to use the term obsessed with seeing the comet. But I was highly interested in seeing something that only flew by the Earth every 76 years.

I enjoyed growing up in the 1960s, being wrapped up in watching science fiction shows, spending hours building models of spaceships, and flying Estes Rockets (something I still do today with my daughter Rachel). Yet, there was something else I was very interested in, too. During the '60s, there was a real-life event called the space race, which was in full force, and the question on everyone's mind was who was going to reach the moon first - the United States or Russia (or at that time as it was called U.S.S.R). It was because of this that I became more fascinated with space. As mentioned earlier, President Kennedy proposed the United States be the first country to reach the moon. Spoiler alert, July 20, 1969: The United States was the first- and to this day, the only country- to land a man on the moon and return him safely. The United States put a total of 12 men on the moon between 1969 and 1972 before NASA stopped all flights there. I envied those who were brave enough to travel into space and walk on the moon.

I vividly remember some of those days of the space race and all that surrounded it back in the 60s. I recall hearing about and then watching some of the later Gemini liftoffs when they were televised, and most of the Apollo rocket launchings. By December of 1968, the Apollo missions started sending flights toward the moon, and how I wished to be able to see the Apollo spacecraft travel to our closest celestial neighbor. I can definitely say that the space race drove my attention to the sky, and my young brain told me there was only one way to look and appreciate the sky, and that would be with a telescope. I didn't know much about telescopes back then, and it would take me years to learn all that went into using them and peering into the night sky, but for Christmas in 1967 or 68, I got my first one.

Do you remember being a child and paging through Christmas catalogs from stores like JC Penney, or Sears Roebuck & Co? I looked for Christmas ideas in 1967 or 68 to give to my parents so they could pass them off to Santa. One item I found was a telescope. Thinking back, I realize it wasn't a very good telescope, but it was the greatest gift ever when I opened it on Christmas morning. It was my gateway to seeing the heavens above.

Much like the one in the photo, this no-frills refractor telescope had a barrel 3 or 4 inches wide and 3 feet long, made out of cardboard and held up by rigid, but not the most durable, aluminum legs. The eyepieces, or oculars that increased the size of whatever you were looking at had plastic lenses (unlike the glass ones made for more professional telescopes). Still, to me, this telescope was the greatest gift Santa ever brought me. Until that Christmas, the closest I had ever been to the skies above was when I was flying a kite.

Needless to say, the telescope had to be assembled and since Christmas Day was too busy, I had to wait a day to put it together. The instructions to assemble it were easy enough, and I was looking through my living room window in no time. The tops of TV antennas across the street during the day were the first things I focused on, but they were upside down when I looked at them. No, it wasn't because I put the telescope together wrong; that is how the inverted image came through, bouncing off the mirror at the bottom end of the tube. However, the telescope did make things look larger, and I couldn't wait for a clear night to try it out.

It was the end of December, and it was cold in Wisconsin. I waited for the first clear night so I could brave the chill and take a deeper look into the night sky. The moon was high that night, and it wasn't even full. Using the plastic oculars that came with my telescope, I did my best to look at the moon's craters; they were the first celestial objects I looked at through my telescope. From then on, a whole different world was above me as my telescope was the key that opened the door into the night sky. I never did see any of the Apollo flights fly to or orbit the moon, as much as I wanted to, but my telescope, no matter how much of a great gift it was, was still severely limited in what it could do. Plus, I was way too much a novice to notice anything spectacular other than the moon. I had no idea where the planets were and didn't realize they were some of the brightest stars in the sky. There was so much to learn about the skies above me and through the following years, I would indeed learn.

Chapter 2

Preparing for Halley's Comet

 Growing up, I was very fascinated with science fiction, space, and the skies above. It had been many years since my first telescope showed me that there was much to see other than dots of white in the night sky. My first telescope lasted a couple of years before the cardboard tube became crushed and unusable. But I never stopped looking up into the night sky. I remember in 1973 when NASA placed Skylab into orbit, and it was during that summer when Skylab flew over my hometown. The time and date it was to pass over Racine was printed in the local paper, and I remember waiting to see it fly over. Fortunately the sky was clear, and Skylab was easy to spot when it passed by. It seemed to fly over in less than 2 minutes, but it was so exciting to see, especially when knowing that astronauts were living in something so high in the sky and orbiting the Earth. I often wondered what kind of view they had of the Earth.

 Now, with Halley's Comet just four or five years away and in the news a little more frequent, I would think of it more often from time to time. However, I needed to figure out what was more important: knowing where to look in the sky when it comes back into view or having the right tools to do so. I decided to get the tools first. I needed a really good telescope; it needed to be much better than my first cardboard tube scope. That one, as impressive as it was to me when I received it, only lasted a few years before becoming unusable. Thus, research or finding something more up-to-date and professional began. I became a subscriber to *Astronomy* and *Sky & Telescope* because both magazines helped tell me what was happening in the sky, including what meteor showers were approaching and what planets would be on the rise in the coming months. But what I found most helpful in these publications were their ads, particularly for telescopes. There was no internet then, and getting information and reviews on buying products was more difficult, so these magazines were extremely beneficial to me in my search.

I settled on a Celestron 8, the 8 referencing the number of inches of the aperture or opening that collects the light of the stars or other deep-space objects. It was not the largest or the smallest telescope on the market, but it best suited my needs. The best price I could find was from the Orion Telescope Center of Santa Cruz, California. With free shipping and no taxes (because it was out of state), it would only cost me $919. But I realized I would have to buy oculars and other things to use, and I would get them as well as other accessories as they became needed.

During the last week of March 1981, I applied for a loan. On April 2, my $1000 loan was approved to pay for the telescope and to help me build a credit rating. By April 3, I had ordered my C-8, an equatorial wedge that allows for polar alignment and is required for long-exposure astrophotography (which I never mastered)-and a tripod for it. I now had my tools to view not just the comet but the rest of the sky above.

A week or so later, I came home from work as a Firefighter just after 7:00 in the morning and found several large boxes waiting for me; my telescope had arrived. I was thrilled and couldn't wait to try it out, but I still had to assemble it and darkness was still 12 hours away. In the coming months, I learned the ins and outs of assembling and using my telescope, and in July of 1981, I ordered star charts and a red lens flashlight so I could read them without interfering with my night vision. I also now had several various size oculars to look not just at the stars and the moon but also at other deep-sky objects like globular clusters, nebulas, and what I considered my favorites, the other planets in our solar system. I was so thrilled when I saw the rings of Saturn and the moons of Jupiter for the first time. I never knew there were so many objects to see in the night sky, and I had been walking under them my entire life!

With Halley's Comet still several years away, I took my telescope out occasionally. I learned there were only certain times of the year when you could see the planets at decent hours (meaning not in the early hours of the morning) due to their and the Earth's orbit around the Sun. I tried to photograph the moon, but the clock drive on the telescope couldn't be adjusted and it was permanently set for deep space objects; I could only take very short time exposures of the moon. I managed to get some decent photos, but nothing like the ones I had seen in the astronomy magazines. I had hoped to get better at it, but never really did. My early photos of the moon were okay and I was still very proud of them, but they could have been better.

By 1984, more articles about Halley's Comet started to pop up in newspapers and magazines like *Time* and *Newsweek*. I even bought a globe of the constellations with Halley's Comet's path and dates as when it would be visible, showing me where and when to look. I displayed it prominently on a shelf in my house, where I saw it daily. I thought it would be my greatest asset in looking for the comet. It would be a wonderful keepsake after the comet finished its trip around the Sun and headed for deep space again. However, in early 1985, I found what I considered the greatest astronomy tool ever for its time.

Home computers were becoming more and more popular in the early 80s, and the one I had was a Commodore 64 that I bought in 1984. It was one of the most up-and-coming personal computers at the time and cornered 40% of that market. It would be considered little by today's standards. Still, it had programs for checkbooks, word processors for writing, and various games, which included flight simulators and Infocom's famous *Zork* series word prose games. One day, while walking through a local Best Buy, I found a program for the Commodore 64 called *Sky Travel: A Window To Our Galaxy*. What first caught my eye was the picture of a telescope on the cover. It was a bright orange telescope that looked a lot like my Celestron-8. I snapped it up, and by the end of the day, I found out how valuable this program would be for any amateur astronomer such as myself.

My Comet Tales

Once loaded into the computer, the program prompts you to enter your latitude and longitude. This was done by loading the map page and moving a 4-point cursor over your or any location you desired. Cross hairs may have been better, but this is what I had to work with. The only disadvantage was looking at in on the old type monitors we used on our computers back in the 80's you had to eyeball your exact location. I did the best I could. After setting you location you would follow up by setting the present date and time, and then it shows you where stars and planets are located in the sky in any direction. This amazing program allowed you to see what the night sky would look like from any place in the world for any given date over a 10,000-year span before and after your present date. Its features also allow you to quickly find the moon, planets, and different constellations. *Sky Travel: A Window To Our Galaxy* could also help you find Halley's Comet, which is just what I needed. However, according to *Sky Travel*, it would only show where the comet was during its possible viewing range from September 3, 1985, through July 31, 1986. This was a nine-month window, but I knew that according to other articles, the best viewing time for the general public without the proper equipment would be less than five months. Another benefit of this program was that it allowed you to print out star charts that could be used outside or in the field to help you find anything from certain stars to deep-sky objects. For me, this was an ideal program and the best tool I could possibly have for allowing me to find Halley's Comet when it returned later in 1985. I cannot say enough about the usefulness and practicality of this program and it was well worth what I paid for it.

The charts I printed and saved from the appearance of Halley's Comet were especially useful to me in the writing of this book. Now armed with my telescope, a computer program that would tell me when and where to look, and time to practice with it, I knew I would be ready to find it when Halley's Comet would come back the following year.

In early 1985, more articles started to pop up about Halley's upcoming approach in the fall, comet fever began to grow, and the countdown officially started for its return. By that summer, another computer program became available; it was called *The Halley Project: A Mission In Our Solar System*.

Considered a space flight simulation game, The Halley Project turned players into space pilots who were given a number of objectives to follow by successfully landing on different planets and moons of our solar system. If you completed all ten levels of the game, you were given a code number to write down on an enclosed card and mail back to Mindscape, the makers of the game. To complete the game, which took me a couple of weeks, I had to find answers to some of the game's questions that directed me to what planet or moon to fly to. To do this, I used a book called *Skyguide: A Field Guide For Amateur Astronomers*.

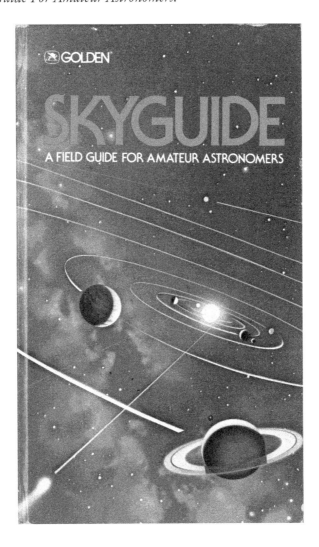

Nick B. Comande

This informative guide was published in 1982 by Western Publishing in my hometown of Racine, Wisconsin. With its help, I completed the game and sent in my code number to Mindscape, and I received a certificate stating that I completed *The Halley Project*. I framed the certificate, but unfortunately, I do not currently know where it is. Playing this game increased my desire to see Halley's Comet, and I could feel the anticipation of its return growing inside me.

Other than trying to perfect my astrophotography, I was ready for the 1985 return of Halley's Comet, which was still several months away. Now, all I had left to do was wait.

Chapter 3

November 20, 1985

**Location: Racine, Wisconsin. 1024 Florence Avenue.
Latitude +42.46 by Longitude -87.46**

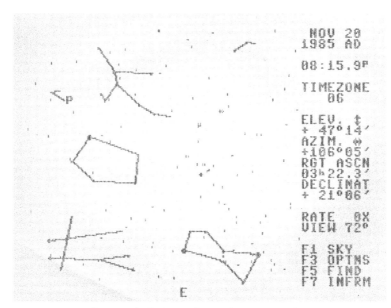

Scan of original Sky Travel star chart. Printed in 1985.

It was while I was walking through my dinette, I noticed that the street light in front of my neighbor's house was out. This occurrence was something that happened from time to time. I actually looked forward to it as it made viewing the night sky with my telescope much easier from my front yard and more of an incentive to take it out. Any light pollution disrupted viewing; even the full or half-full moon washed away the smaller stars and hampered what you could see with and without a telescope, but there was a moon tonight, not full but in the South sky away from the direction I would be looking. With that light out, it made looking at stars, planets, and even some deep-space objects possible. Not optimal, but possible. I walked outside to look up at the sky to see if I could see stars, meaning it was clear out. It hadn't been for some time now, as there had been many nights where the sky was filled with clouds. I decided to take out my telescope and try to find Halley's Comet. I had been waiting years for this, and I wondered if tonight would be that night, even while viewing in the city. There was only one way to find out for sure: to try and see for myself. Now, with more mentions about the comet on the news at night and various telescopes worldwide had started tracking Halley's Comet, I had to try.

I first looked up the location of the comet with my *Sky Travel* program. According to the program, the comet would be small and have no visible tail (yet), but I was well within the viewing dates it had listed on its program. I mentioned it before, and I will again, saying this program was worth every cent I paid for it and more, especially in the early and final viewing days of the comet. The program showed me that the comet would be in the Eastern sky, an ideal place for me, and it confirmed that the moon was closer to the South Western sky, the opposite direction I'd be looking. With houses all around me, the higher up in the sky, the better it was for me to view anything. By 8:15 pm, I had my telescope set up, and impatiently I wanted to start my search. Thinking back, I would pick out a planet or other deep-sky object and try to find it. I remember how thrilled I was when, for the first time, I found the Ring Nebula or a distant Globular Cluster of stars. Halley's Comet would be my greatest find to date, at least in my eyes.

I was always careful not to rush in setting up the telescope or even when I was carrying it outside. It would have been an awful thing for me to have an accident, to trip and fall, break it, and that would bring a quick end to my years-long wait; that would be nothing short of devastating to me. It would have been a terrible thing to have waited for years to see something and then lose my means of seeing it. While I was trying not to rush to set it up in the street in front of my house, I was careful when placing the legs of the tripod in the street that I had marked years ago. These spots, along with the equatorial wedge, kept the telescope in line with the stars while the Earth spun on its axis. This was important when trying to take timed photos. I could feel my heart racing the whole time, knowing that this could be the day, or rather night, I was to fulfill a dream. Maybe dream is the wrong word. I didn't want to say quest, and though it might be considered a bucket list item, this goal of mine was more of a need or deep desire. Regardless, my years of waiting might finally be coming to an end this evening.

My target was in the Eastern sky, above the constellation Orion. This was one of my personal favorites; I even have a star named after me in it. With the comet craze building over the past years, more people were looking up and again noticing the night sky. There was a place where you could buy a star and were given the coordinates for it. Mind you, these were tiny stars, and you did need the means to find them, such as a large-diameter telescope like mine, where you could search for things using the right accession and declination settings. It also came with a fancy certificate. I often wondered how many people were given the same star. But the certificate was impressive to look at; I framed mine. In 2020, my neighbor bought me a piece of the planet Mars in the same type of promotion brought about due to the recent landing of probes there. With the coming of the comet and more and more stories about it in the paper and magazines once again, people's interest in the sky was brought back.

Impatiently with my heart racing, I started with my largest eyepiece (40mm) to cover a larger area. If I found my target, I would then use a higher-power lens, but it decreases the size of the viewing area. The sky was beautiful that night; the air was clear and cold. Visibility was good, even with the moon hanging low in the Southern sky. It wasn't full, but it was getting close. The brighter the moon, the more light that was being reflected towards the Earth, making the night skies not so dark; at least, it was in a different direction that I was looking. It wouldn't affect viewing for a while yet, so I was thankful. When looking for something that was roughly ten by five miles in size in what I consider an infinite sky above me, I would need all the help I could get.

Anticipation had been growing more and more as I learned that the comet was approaching the Earth, and my heart would sometimes race knowing that it was within viewing distance. To think I might actually see Halley's Comet tonight. I had waited so long for this, and I didn't know what I really expected to see what it would look like or if I would know it if I saw it. There were photos in magazines taken by telescopes much larger than mine that gave me an idea of what to expect. I hoped to recognize it when I first saw it, and you know what, I did.

It was small and fuzzy, not unlike the descriptions that were in the media. I wasn't 100% sure, but I thought I saw a slight or ever so faint orange tint to it, perhaps reflected light from the sun; I didn't know for sure. There was no visible tail on it yet because the comet was so far from the Earth and the sun; the tail of the comet would still be behind it and out of view I knew that as the comet became closer to the sun, the solar winds and the pressure of the sunlight on cosmic dust would cause its tail to grow millions of mile long, but that was still months away. It almost seemed out of place, and with the light pollution from the city, I would not be able to see the tail for some time yet. Even if I weren't in the city with darker skies, I would not have been able to see the comet's tail tonight.

But what I did see thrilled me to no end. After years of waiting, there it was. It was not as impressive looking as I had seen in the photos taken of it back in 1910, but it was there and coming back to visit. According to an article I read in my local paper back in July of this year, Halley's Comet speed had increase from it 2000 miles per hour from it farthest distance to the sun to about 48,000 miles per hour in mid-June of this year. The comet would increase in speed as it became closer to the sun and would eventually reach a speed of over 120,000 miles per hour. It was reported to be 430 million miles from the Earth in Mid-June of 1985.

I considered this the most incredible thing I ever found in the night sky. I was so excited when I saw it; I couldn't help but give out a loud yell to say I found it. When speaking with my neighbor the next day and telling him of my discovery, he said he heard me yell. With the clock drive of my telescope turned on to slightly help counteract the rotation of the Earth, I didn't have to keep repositioning it to keep the comet in my field of view as often, but the clock drive is perpetually dedicated to view stars or deep sky objects. I didn't have the other electronics to speed it up or slow it down for planet or comet viewing. I enthusiastically replaced the 40 mm eyepiece or ocular with my 25mm. This would enlarge the object I was looking at, and if I did it quickly enough, it would not have moved too far from my field of view, and I would be able to bring it into focus faster. What I saw was just a larger version of what I saw before, but just as thrilling and just as exciting. I couldn't believe it, the great Comet Halley, in my view in the night sky. For roughly the past 4 years, I had been waiting for this, and today was the day a dream (or maybe a quest) came true. I'm not sure if it was because of the joy I was feeling or the coldness of the night air, but I wiped a tear from my cheek. I knew in the months to come, as it changed its position in the sky and became closer to the sun, I would see the great comet's tail, and for that, I would have to wait just a little longer.

At about 8:35 pm, a friend of mine from the fire department was on his way to work. His name was Rick Bowman, and he happened to look down my block as he drove by and saw me standing in the middle of the street. (Luckily, there was very little traffic on my street). He swung around and asked what I was doing. I told him that I just found Halley's Comet and offered him a look. Rick was thrilled as he had never looked through a telescope like mine, but he was the first person I showed the comet to. Rick was thrilled to see Halley's Comet but said, "After seeing such a once-in-a-lifetime sight, everything else I ever see through a telescope will be second class." Needless to say, it was still an exciting evening, not just for Rick but especially for me. He was my witness to this celestial find, the great comet Halley. Something I learned about many years ago, something that I waited, planned, and prepared for, was finally in my grasp. Years of waiting, an infinite sky, a good telescope, and a computer program from Best Buy all helped to make one of my dreams come true. But it wasn't over yet; the comet was still far away and slowly getting closer at almost 50,000 miles per hour, and the best was yet to come.

Chapter 4

December 6, 1985

**Location: Racine, Wisconsin. 1024 Florence Avenue.
Latitude +42.46 by Longitude -87.46
No star chart for this day**

It was just over a week since I saw the great Comet Halley. The weather has not been kind, and there have been more clouds in the sky than there were stars. In the past few days, I was thrilled to tell my family and co-workers about last week's celestial find, even if it was not as impressive looking as I had hoped. But I wanted, or needed to take another look and see how much it may have changed in size and if I could see its tail or not. Even my *Sky Travel* program had yet to show a tail on it, but it should start to grow the closer it gets to the sun.

It was getting colder out with winter approaching, and there were more intermittent clouds in the sky, even a light dusting of snow a couple of days ago. I carried my telescope to the street where I found the comet for the first time. It would still be high enough in the sky that I might see it from here again. Finding it would be easier now because I knew exactly what I was looking for. But it just wasn't in the cards for me tonight; with fast-moving clouds above me, I quickly set up the telescope to try to beat the weather, but I wasn't quick enough. Though I had the telescope pointing in the right direction and thought I may have seen Halley through my finder scope, the clouds came in too fast, too thick, and with no end in sight. So once again, I would be forced to wait, so close and yet so far, still millions of miles away with a thin layer of vapor between me and the comet.

With some disappointment, I carried the telescope back into the house. On November 27th, the comet was to be closest to the Earth while it is on this side of the sun, being a mere 58 million miles away. It will be 20 million miles closer when I will try to view it in late March-April 1986. I was hoping for my second glimpse of the comet but was not to be. I still had several months in which to see the comet, and this was just the beginning of its journey into Earth's neighborhood.

Chapter 5

January 11, 1986

Location: Racine, Wisconsin. The Western side of Racine. West of Latitude +42.46 by Longitude -87.46 (Roughly)

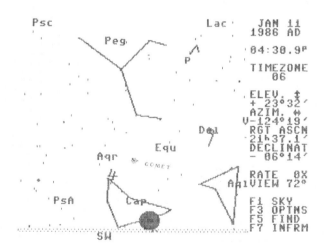

Scan of original Sky Travel star chart. Printed in 1986.

It had been over a month since I first saw Halley's Comet. By pure misfortune, I had to work every night that the sky was clear and on my off days if it was extremely cloudy, there was just enough overcast to prevent viewing. It was hard to believe that already one-tenth of my six month viewing window was over. My job as a firefighter required me to work a 24-hour shift, meaning I was away from home all night. One time, I brought my telescope to work and placed it on top of the Safety Building downtown—the main headquarters for the city's fire and police station—but the terrible light pollution and cloudy skies kept me from finding the comet from there.

Since my last sighting of the comet, it had traveled from the Southeastern sky to the West-Southwest sky as its speed constantly increased as it neared the sun. My front yard or the middle of my street was no longer a good place to see the comet from as it appeared low in the sky and followed the sun as it set. Viewing time would now be anywhere from one-and-a-half to two hours, depending on possible obstructions on the horizon. However, according to *Sky Travel*, the comet would be easy to find so I would not even have to use the setting circles on my telescope.

Because the comet had been traveling thousands of miles closer to the sun, increasing in speed along the way, it would also be brighter and more prominent than the last time I saw it—the comet might even be in a position where its tail would begin to be visible. But, to see the comet and its tail, I would have to get away from the city's lights and escape to somewhere with a view of the Western sky and electricity to power the clock drive of my telescope. So, I packed up my telescope and went to my sister Pat's house. Pat lived with her husband, Tom, and their two children, Andrew and Benjamin. They lived just West of the city, not far from me, but they had a better view of the Western sky with fewer obstructions. Spotting Halley's Comet would be much easier from their location, and I would not have to contend with rows of trees or houses like I did on my street.

By 4:30 pm, I had set up in their driveway and Tom helped run an extension cord from his garage to my telescope. I do not remember what the exact temperature was, but it was cold—at least in the low teens. The sun would set by 5:00 or 5:15 pm, and the comet would follow shortly after that. I saw the comet easily from Pat's house. It was just a few degrees above Jupiter and a hint to the right. Jupiter was, and had always been, an easy planet to find, even with the sun still setting. I found Halley's Comet with no problem just by looking through my finder scope. It was much easier this time than when I first located it for the first time back in November.

This was also my first daytime viewing of the comet, with the sun still barely above the horizon.

Through the finder scope, I saw a brighter star that seemed out of place and it had a slightly different shape to it. Right away it told me that it was not a star and had to be Halley's Comet. As I looked at the comet through my telescope, it still appeared as a fuzzy cotton ball, but noticeably larger and more oblong than when I first found it well over a month ago—it had at least doubled in size! Straining my eyes, I looked to see the tail, but there was just too much light in the sky from the setting sun. Perhaps with some short-time exposure pictures when the comet got closer the tail would show up in them. Because of the angle, my telescope only needed to be corrected every few minutes. This meant the comet either had little movement or it was working just right with the Earth's rotation, resulting in a less blurry image on film. (Little did I know then that I still had much to learn about astrophotography.) After the sun had set, the planet Jupiter was much brighter and a very easy find in the Western sky. You could now see the comet without any real detail with the naked eye, provided you knew where to look and were a little familiar with the sky. Halley's Comet was just above the constellation Capricorn which was not far from the moon. Though it was just a sliver—a fingernail clipping of God, as some would say—it still meant just a minimum of reflected light, and all light was terrible if you wanted to see the tail of the comet.

With great enthusiasm, I called Pat and Tom from their house. They had previously expressed interest in seeing the comet, so I knew they would like to see it. Pat said she felt special because she was one of those lucky enough to see the comet through a telescope. With the great comet finally in viewing range, the media had been hyping its return much more by now, and "Halley Fever" was starting to hit the general public. The desire to see it first hit me big around five years ago, long before most people were even aware of its return. Pat and Tom saw the comet and were impressed, but what I really wanted to do was show the comet to my nephews, Andrew and Benjamin. They were both just under 10 years old. I would also have to show the comet to my other nephew, Ed, who was eleven and the only child of my oldest sister Sue and her husband Gary.

I thought to myself, because of their young age, they at least have the chance of seeing this spectacle again in 2061 when Halley's Comet returns. As for myself, I am pretty sure I will be a one-time viewer. I did not know anyone who has seen the comet before, and there can't be many people around who have seen it twice in a lifetime (and remember it), but my nephews have a better chance than me.

With my telescope set up, now with the clock drive now running and again focused in on the comet, I let my nephews have a look. They each took a turn, and I made certain that they saw Halley's Comet because I had them describe it to me. I let them look as long as they wanted and hoped they would remember what they had seen. But would they remember enough to compare it to the comet when it returned 75 years from now? Who knows? I asked them at the time to write down what they saw or even draw a picture of the comet, but I do not believe either did.

I questioned them that night about what they thought about the comet. Andrew said he was amazed because, "It was a real experience" and that he "would like to see it again" when it returned in 75-76 years. As of 2061, Andrew and Benjamin would be in their low 80's. Benjamin reminisced about the comet appearing as a fuzzy, white ball and—although he did not know when it was set to return or anything else about Halley's Comet—he was glad to have seen it and hoped to set eyes on it again.

I am sure I will be long gone before the comet swings by this neighborhood again, but they have a chance to spot it in 2061, and I hope they do. Regardless, I will always be grateful that I was able to be part of their first viewing. Maybe when they see Halley's Comet again, they will remember this night and what their uncle did for them.

Earlier I mentioned that viewing the comet tonight was easy, visually that is. But the viewing conditions were not. The cold temperature—which I learned from Tom was 15º— made working with the telescope difficult. It was very cold to the touch which sometimes made it hard to focus and realign. Plus, doing so without gloves to have better control of the knobs for fine-tuning the focus froze my fingers. Whenever the wind blew, it was just another reminder of how cold it really was outside, and the strong wind would shake the telescope when I looked through it. All this will make trying to take a photo more difficult, but before the comet followed the sun below the horizon and after everyone had multiple times to view it and get their fill, I quickly hooked up my camera adapter to the telescope. This was my other goal of the night, to get a picture of the great comet.

This picture did not necessarily have to be a good one—though that would be nice. It just had to say, I saw the great Halley's Comet with my own telescope and here was the photo that proved it. (Unfortunately, I did not get a viewable photo this night.)

The comet was getting low in the sky, and I was cold, even though I had dressed warmly. It was difficult to stay warm as I just stood around and looked through the telescope. I decided to say goodbye to the comet for the night and then packed everything up. The comet was down by 7:00 pm. But, I would be back in two days since I have to work tomorrow, before I could see it again—weather pending, of course—on January 13th. The moon would be shining a little brighter as it grew, which was always a drawback, but I plan to make the best of it. The comet was finally here and, now with the clock now running until it was gone, I had to make the most of each night if possible.

Chapter 6

January 13, 1986

Location: Racine, Wisconsin. Western side of Racine. West of Latitude +42.46 by Longitude -87.46 (Roughly)

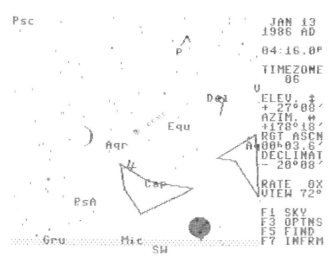

Scan of original Sky Travel star chart. Printed in 1986.

There was not much difference in the location of the comet this evening and that of two nights ago. It was still in the West-Southwest sky, below Pegasus and above Jupiter and the constellation Capricorn. I was set up again at my sister Pat's house, early enough to view the comet in the daylight. This meant that I would lose sight of the comet as it fell behind the house across the street, but I thought it wiser in the long run. I was all set up by 4:10pm. On this night I was interested in photographing the comet and the real trick was to be ready to view or photograph the comet right after the sun went below the horizon and before the comet did the same. I set up behind my sister's house this time so light from the headlights of passing cars would not interfere with the photos. With the help of my nephew Andrew, we attempted to photograph Halley's Comet. I would not know if the pictures would turn out for a while though.

Until now, I had only been successful in photographing the moon, and those pictures weren't very good. I asked some photographers about trying to get a decent picture of the comet with a telescope, but they were not of much help to me as they never taken photos through a telescope before. I thought it shouldn't be much different than using a very large telephoto lens, but it was more difficult. When trying to photograph the moon, I would try different camera settings and time exposures. Unlike today with digital cameras, I had to wait until I filled a roll of film, take it to someplace to get developed, wait a couple of days to get it back and then hope the photos were in order with the negatives so I could compare it to my camera settings which I didn't always write down making this a tedious and costly venture.

The winter winds were constantly shaking the telescope, so much worse that the nights before. This too would work against my attempts as I tried to take photos. While there were less trees in the area and the ground more level, there was nothing to block the wind. The cold was hard on the fingers when setting up the camera; and I worried about the affect it may have on the clock drive mechanism. But I tried. The moon was a slightly larger tonight, though still a crescent, it cast light, while it might not be much, it would hurt. The sky was very clear which was great for viewing and the temperature was again very cold. The comet suddenly appeared to the naked eye if you knew where to look, and all who had seen it on the 11th had seen it again tonight. Some neighbors from across the street had come over about 6:00pm that evening, but the comet was already down for me, actually just out of site from my stand point. Not really down for the evening, but blocked from view at my sister's house. By the time it would have taken me to me to set up the scope in a more favorable position, the comet would certainly have been down below the horizon for the night. I gave assurance to all that if the weather was clear tomorrow night I would surely be back. And I was. I did not get a photo this night as well that showed the comet.

Chapter 7

January 14, 1986

Location: Racine, Wisconsin. Western side of Racine. West of Latitude +42.46 by Longitude -87.46 (Roughly)

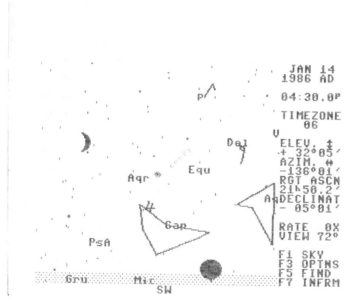

Scan of original Sky Travel star chart. Printed in 1986.

It was Tuesday, January 14, and, as promised, I returned to my sister Pat's house in West Racine. The sky was clear and the air was cold. I set my telescope up in the driveway in front of the house—close to the same spot where we had viewed it from the previous nights—and by 4:30 pm,

My Comet Tales

I was ready to start looking for Halley's Comet. The sun was still setting, and the moon was slowly becoming a quarter full.

My sister Pat called her friends and neighbors who had missed the comet last night. They, in turn, reached out to some of their neighbors who also came to look at the comet. When the sun had set at 5:15 pm, a dozen people were in her driveway looking at the comet through my telescope and the one pair of binoculars someone had brought. Even though it was cold out, Pat suggested we celebrate the viewing of Halley's Comet—so we did. This was the first official comet party I had the pleasure of attending. Pat went into her house and returned with a bottle of champagne we drank from plastic cups to celebrate the return of the comet. We knew it was freezing out because ice crystals were forming in the cups. Some people brought warm beverages to keep warm while waiting a turn to look through my telescope. As cold as it was, we were having a great time looking at and talking about Halley's Comet. That was when the police showed up.

Nobody called the police; the officer was just making rounds and wondered what all these people were doing, standing in a driveway and drinking in such cold weather. When he pulled up and asked, I told him. Halley's Comet was in the news every other night now, and even if you were not interested in it, you at least knew about it. It didn't take long to explain what we were doing to the officer, especially with my telescope standing tall in the driveway behind me. I offered him a look through it, which he quickly took me up on. Not everyone had a telescope as good and as powerful as mine, so the officer was impressed and glad he had been able to see the comet. The officer said, "The comet looked like a fuzzy snowball, " which was—funny enough—the exact description everyone else gave when they saw the comet for the first time, either tonight or on previous nights. The police officer stayed a short while, took several more looks, and then left to continue his rounds after having thanked me for letting him see Halley's Comet.

The cold took its toll on everyone in my sister's driveway. With nothing to block the wind, you felt more of a chill when it blew, and it was even worse with the sun now set. I was surprised that my sister's neighbors lasted as long as they did, but this was a once-in-a-lifetime experience for them and, like me, they did not want it to end. One by one, everyone took turns looking at the great comet and when people were waiting to take another look at it they were talking about they had heard in the news as the media was now following it more closely. The media also mentioned that spacecraft launched from several nations would soon be making their approach to the comet. Perhaps they would take photos displaying the comet's great tail and show them on TV. The mission of one of these craft is to measure the change in the size of the comet as the solar winds burn off its surface to create its immense tail. It may tell us how many more trips Halley will make around our sun before it depletes itself and is no longer, ending thousands of years of travel between our sun and deep space.

Halley's Comet was still easy to find, even without seeing a long tail behind it, and it was still a thrill to see. To me and the others, this was a historic event that we were taking part of. When I was asked why I bought my telescope, I said it was originally to see Halley's Comet, but there was still a lot of other things that I could look at in the night sky.

I showed the planet Jupiter to the crowd and they were amazed to see it and four of its moons, it too was quickly moving out of viewing range and it dropped below the horizon by 6:30 pm I then showed those interested one of my favorite constellations, Orion, was high in a very dark Southern sky, and in it was the Orion Nebula. It is such a beautiful thing to see on a clear night.

By now, the comet was very low in the sky and it was about to go below the horizon and out of sight for today. Once that happened, comet viewing for the night would end, so I knew I had only minutes left to show the comet to those who wanted to see it, perhaps for them a last time.

The comet party, although short and sweet, began to break up once Halley's Comet was no longer visible. Everyone took a last look at it before leaving, knowing they would never see Halley's Comet again. I felt they were satisfied when they left and I could see them as they walked away, smiling and talking about the comet— a part of history that had come once during their lifetime. I felt good inside, knowing I brought them something no one else could. If the day came when I could show the wonders of the cosmos to someone else, like I did tonight, I knew I would be more than happy to do it again.

Chapter 8

January and February 1986

By the end of January 1986, the great comet was extremely low in the Western sky and being able to view it was sometimes not worth the effort of trying to find a place with a very good exposure of the sky with no obstructions to be able to see it from my city. I admit, I was beginning to get tired of packing up my telescope in its case and taking it out and reassembling it each time I took it somewhere and the cold sometimes made it less enjoyable. But looking back now, it was a small price to pay for fulfilling my dream. I left it all assembled when I was at home and was able to carry the whole thing outside when I wanted to use it in front of my house, but there were too many trees and houses by me to see the comet. It was still a good spot for looking at objects higher in the sky, but not for things near the horizon. My sister Pat's house was no longer a prime viewing spot, but I was grateful for the nights I had there and I'm sure her family and neighbors were appreciative as well.

The viewing window was closing and soon Halley would be below the horizon for over a month, at least in my part of the world. There were cloudy nights opposite of the days I worked which again kept me from seeing it any more in the month of January, and I would not see it again until after it orbited around the sun. The Earth was moving away from the comet at this time, but as Halley moved closer to the sun, it would increase its speed and the comets tail would grow, but I stood less of a chance of seeing it now. It is during this time Halley's Comet would makes its closest approach to the sun, this is called perihelion which would be on February 9, 1986. It will be about 56 million miles away from the sun. Mercury, the sun's closest neighbor's elliptical orbit is between 29 and 43 million miles away from the sun. My next best chance to see the comet would be in March of 1986, when it would be heading back near the Earth rising higher in the sky making its closest approach on April 11th 1986, but even then, it would still be low in the sky. I should also point out that the comet would be leaving this part of the solar system much faster than during its first approach to the Earth, that being due to what I believe is called the sling shot affect around the sun. The comet was now traveling at over 50,000 miles per hour at perihelion picking up speed to leave this part of the solar system at 122,000 miles per hour.

I considered myself very fortunate as Halley's Comet rounded the sun on this particular journey, it came within viewing distance of the Earth twice. This does not always happen, but this time it did. I didn't know about its past visits, but I was glad that it would. Looking at a drawing in my local paper showed how the comet would pass by the Earth in 1985 and again on the other side of the sun months later and much closer to the Earth in 1986. Making this visit a special one for me with more chances to see it. This doesn't always happen and it will not happen on its next visit in 2061. So, I was extremely fortunate to have not one but two viewings of the comets passing.

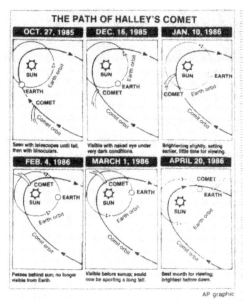

Clipping from Racine Journal Times. Dec 6, 1985 (Source NASA)

This illustration in the Racine Journal Times (supplied by NASA) showed how Halley's Comet is moving around the sun in a clockwise orbit where the Earth is traveling in a counter clockwise orbit. I consider myself extremely fortunate that the Earth is in the right place to cross paths with the comet not once, but twice on this particular visit. I wonder how many times that has happened in its past visits? The disadvantage to this particular passing was the fact that when Halley's Comet was closest to the sun, sporting a most magnificent tale, the Earth will be on opposite side of the sun robbing us of it most spectacular view.

While I would miss seeing Halley's Comet while in February entirely, the coldest month in Wisconsin where the cold makes the air clearer for viewing, it also makes working with the telescope more difficult while using gloves, or not using them and freezing your fingers which is the other side of the proverbial coin in winter astronomy.

I was thankful that when I was able to first find the comet, it was in the early evening hours and high in the sky. When the comet comes back, my best viewing times would be in the early hours of the morning, an inconvenience I would have to deal with if I wanted to see the comet in all its glory with a long tail behind it. When Halley makes its closest approach to the Earth during this pass, the best place to see it would be in in the Southern hemisphere, Australia to be exact. I wouldn't have the ability to see it there, so I would have to make do with what I can see from here. According to my *Sky Travel* program, my optimum viewing direction would be in the East-Southeast sky. Fortunately living right next to Lake Michigan, there would be no light pollution in the area I would be looking. There will be residual light from the Chicago area, but I hoped to not be looking that far South, at least not at first.

There was something else that kept my mind and eyes on the cosmos was when I was young and would stay up late on weekends. I would flip through the channels on my TV, and remember, there weren't the hundreds of channels that there are now, I ran across an interesting feature on public television. It was a weekly 5-minute show that came on just before or after midnight and it was called Jack Horkhiemer: *Star Hustler*. Horkheimer who was originally from my home state of Wisconsin was the director of Miami Space Transit Planetarium in Florida. His show started in 1976, but I did not learn about it until it went national in 1985. In 1997, the shows name was changed to Jack Horkhiemer: *Star Gazer*. His weekly shows pointed out different things in the sky's above, such as when planets would align, where to look for meteor showers, when comets were coming by (Halley's was not the only one, but the most popular). He would tell you when and where to look in the sky, giving back yard astronomers tips and insight on the night sky. I know that I followed his advice a number of times over the years. His show ran from 1976 to 2010, and I still remember his final words at the end of each show "Keep Looking Up", it is a practice I still do today.

By the end of January 1986 comet fever was growing more and more as the best views of the comet were approaching in the coming months. The further South you were the bigger the hype. As earlier mentioned, the Southern hemisphere was the place to be. Florida which being more than a thousand miles South of me had better viewing than I would have as the comet would be slightly higher in the sky. At that time Florida had 28 small observatories offering viewing packages to view the comet. While there were others who also had a desire to see the comet, they too looked for ways to view it. One of the more popular ways was to take a cruise. Regular cruises that sailed through the Caribbean were now called Comet Cruises. They didn't change their routes or anything, just changed the name of their cruise. It was actually a great marketing ploy and gave anyone on them a good chance to see the comet.

By the very end of February and early March, Halley's Comet would be now considered a naked eye event. Meaning you didn't need a telescope or binoculars to see it. Depending how far away from city lights you were, you might even see part of its tail. A cruise ship out at sea, far away from all the light pollution of the cities would make a great place for viewing it so I decided to look into some of the cruises for myself.

With the help of my trusty *Sky Travel* program, I looked into a few cruises at my local travel agency. Remember the internet back in the mid 80's for looking things up was not like it is now, there was no google until the late 1990's. I took brochures from the travel agency I dealt with at the time to look at the different routes the various cruise lines were taking in order to see which ones were traveling the furthest South and figured out which might have the most and best viewing areas at night or early morning as the case may be. I added the latitude and longitude coordinates to my *Sky Travel* program and searched Halley's Comet to find out where it would be in the sky. One cruise line, Costa had a cruise that I could afford and went as far south as Caracas, Venezuela. It was not the Southern hemisphere, but it would be the best that I could do, so I decided to give it some thought. After all, the comet traveled millions of miles to swing past the Earth. I could travel a couple of thousand miles to see it better.

I decided to take the cruise. One of the reasons that made me do it was because work had been rough lately. As a Firefighter and EMT (Emergency Medical Technician) I spent a lot of time on a rescue squad. My goal was to become a paramedic as soon as the department started a paramedic program, little did I know that that was still over ten years away before my city would have paramedics. You see all kinds of things while working on the back of a rescue squad. Some good, some bad, sometime there was stuff that you wish you didn't see and to this day I still have memories that occasionally haunt my dreams at night. During the early part of 1986, I had a 5-week period with a number of SIDS deaths (Sudden Infant Death Syndrome). There were many hard things I had to deal with on a rescue squad, but telling a young couple that their infant child had passed was one of the worst. It was situations like that I couldn't erase from my mind. It weighs on you after a while, at least it did me. I think that was one of the things that pushed me into a cruise, I needed to get away and since I was into Halley's Comet and waited years to see it, I thought, why not, I will only see it once in my lifetime, so I decided to make the best of it and off to Caribbean I went.

I did make the decision to bring my telescope, hoping to get some good photos, at least one. I looked at the route that my ship was taking and estimated where it would be before making the different ports. I adjusted my *Sky Travel* program map to the different latitudes and longitudes for each morning as that is when the comet would be best viewed and printed up the star charts showing where the comet would be to take along. These are the very same papers that I'm using in this book. On March 15, 1986, with my *Sky Travel* charts in hand, I flew to San Juan, Puerto Rico to board my comet cruise ship.

Chapter 9

March 16, 1986

Location: South Caribbean Sea
Latitude +13.11 by Longitude by -70.52

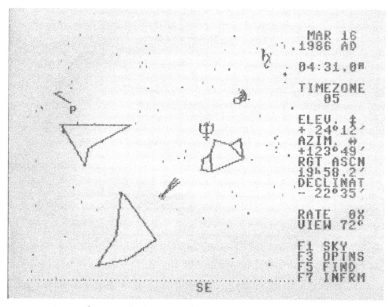

Scan of original Sky Travel star chart. Printed in 1986.

Tonight—or I should say this morning—would be an extremely different viewing of Halley's Comet. It was 4:00 am, and the latitude and longitude were very different from those of my previous comet encounters. I was on a cruise ship in the Caribbean, at a close approximation of 13° latitude by 70° longitude, in a place where I would not have to worry about light pollution. There were no city lights, houses, street lights—nothing. Since the ship had radar, it could turn off all its deck lights, save its running or mandatory navigation lights. It was so dark that I stumbled up an outside stairway because I couldn't see the stairs. Once I arrived on the observation deck, I found myself alone. Not a soul was there, but why, I wondered. The deck's emptiness surprised me since I had talked to several people who said they would get up to see the comet. Admittedly, I had no trouble getting up because I never went to bed. Adrenaline surged through me as I waited eagerly to see Halley's Comet as I had never seen it before.

As I planned for this cruise, I spent a fair amount of time studying star charts and working with the *Sky Travel* program to become familiar with the stars in the South Caribbean Sea's sky. I even brought copies of the Sky Travel charts to help me if I needed them. (These are the same charts I have used to help me write this book, but now—since they are now over 35 years old—they are faded and yellowed.)

As soon as my eyes adjusted to the darkness, I found a place to stand by the rail of the forward observation deck. I purposely did not look up into the sky for the comet until my eyes were well-adjusted to the darkness. The wind was so strong that it felt almost cold on this cruise ship in the Caribbean. This was a distinct change in temperature from what it felt like 12 hours ago as I stood on the deck in the warm sun. The ship was cruising faster now than when it first left port, and I noticed the ship's roll and pitch for the first time. (Needless to say, the horror quickly set in.) Having never been on a cruise ship before, I had not realized the ship would be constantly moving—shifting up and down with the swells of the waves—making any time exposure photography impossible. Even trying to keep the comet in the center of the telescope's viewing area for simple observation would be challenging. I recognized that I would have been better off bringing a good pair of binoculars to see the comet, but I wanted to take photos—a feat that would now be next to impossible. This was, admittedly, a poor start for this comet cruise of mine. I traveled over 2000 miles South with my telescope, only to find that it would be nearly useless.

The sky was unlike one I had ever seen before. It was certainly darker than those of northern Wisconsin. Even without using my telescope, I was more than amazed at what I saw. The sky to the north looked extremely different than what I was used to looking at as it was filled with more stars than I had ever imagined. I knew I would see more stars than usual just because of the lack of light pollution, but I never imagined a sky like this that included clusters of stars so big that they could have been mistaken for clouds. I was awed at this sight; words could not describe the beauty I saw in this night sky. Like the comet, it was something that had to be experienced.

Since I was facing north, I thought I should be able to see the comet if I turned around. It had been roughly six weeks since my last viewing of Halley as it rounded behind the sun and out of sight. I was eager not only to see Halley's Comet again but also to observe any changes since my last viewing in January. Halley was moving farther away from the Earth toward the sun at that time. Now, it was coming closer to the Earth again than when I first viewed it. I prepared myself for this moment as I turned slowly toward the direction of the comet, only to see what every astronomer fears. The entire Southeastern sky was filled with a solid line of clouds. My hopes of seeing the comet this early morning, since it was after 4:00 am, sank faster than a rock in the water.

I wrote an entry for this day—even though I did not see the comet—because there is still a great deal to see in these Southern skies. I was trying to make the best of a bad situation by seeing some of the other sights in the rest of the sky, which were not covered by clouds in the Southwestern sky. The constellation Crux, more commonly known as the Southern Cross, was hanging in the Southwest sky, barely above the horizon and not too far from the cloud line blocking my view of the comet. The Southern Cross, as told in legend, showed sailors the way sailing from Europe to Australia. There was also the popular song of the same name released in 1982 (by Crosby, Stills, and Nash) that told about sailing in the Southern seas and using the stars for navigation. Even the flag of New Zealand has the stars of the Southern Cross on it. It was a beautiful sight and something not readily visible in the northern Midwest. I thought of all the sailors who stared at it or used it to navigate their ships over the past centuries and was happy to be—at least momentarily—in its company. Saturn was due South of me at this time—right on the edge of the cloud line—visible, but without my telescope, I would not see its mighty rings.

It was still a big disappointment not to see Halley's Comet today. I had gone for weeks, unable to see the comet back home due to either weather conditions or its location behind the sun, and I lost my first night of viewing. Now, I only had seven nights left to see the comet while in the Caribbean, and I hoped this would be the only time the clouds would work against me.

Chapter 10

March 17, 1986

Location: Caribbean Sea, Northeast of Curacao
Latitude +12.21 by Longitude -69.00

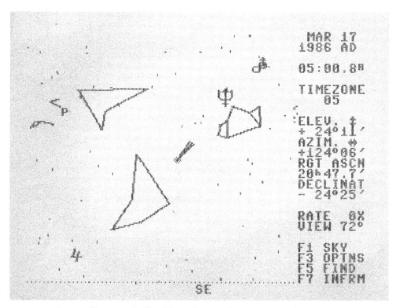

Scan of original Sky Travel star chart. Printed in 1986.

It was 4:00 am on board my ship, the Costa Carla. It was time to return to the observation deck toward the ship's bow. I had been checking the sky periodically, hoping not to see clouds roll in, but fortunately, there were stars in all directions. This time, I waited a bit before going up on deck so my eyes would adjust to the darkness and I could catch the comet in full view. I decided not to watch the comet grow as it came over the horizon; I wanted to see it in its full glory. Once again, I made my way up to the pitch-black observation deck, and for the second time, I was the only person up there. For ten minutes, I looked toward the north, away from the comet, so my eyes could adjust even more to the darkness. I'm unsure if it was a full ten minutes or not, it seemed like an eternity and I held off as long as I could before looking to the South. I turned around slowly, taking a quick glimpse of the Southern Cross in the Southwest sky, so low on the horizon, while noticing no clouds in the sky. I didn't know how long it would take me to find the comet. I was going to start my search in the sky by finding the constellation Sagittarius and looking East of that, but I really didn't have to search all that hard.

There it was, hanging in the Southeast sky, somewhat higher above the horizon than I had anticipated and more elevated than my *Sky Travel* program led me to believe. I couldn't help but see it. Halley's Comet was just as bright as the planet Jupiter was when I viewed it back in Wisconsin and almost as big. The tail of the comet was brighter and more visible than I had hoped, more than the length of a full moon. It was reaching toward the constellation Sagittarius. The area of the tail next to the comet was so bright that it was opaque and became more transparent and sparser the further it reached away from the comet. I could see stars through the tail, but I couldn't tell for sure without a telescope or binoculars. I was so thrilled to finally see the comet again; this time, it truly did look like a comet, tail and all. It was so much more beautiful and exciting to see, reminding me of that night in November of last year when I first found it in the night sky. This was the very moment I had waited years for. With my goal finally fulfilled, I had genuinely seen the great Halley's Comet. What had been years in the making was here, and with the exception of viewing it in the next few days to follow, I will never see the comet again as well as I did at this particular time. As I stood on the deck alone, still taken in by what was before me, I didn't even feel the strong, cool winds blowing over me. I was oblivious to everything around me, yet I was disappointed that there was no one else there who cared to see it. While it was nice having the whole observation deck to myself and finding a place out of the wind to view the comet as the ship slowly rose and sank between the swells as we traveled the Caribbean Sea, it seemed a bit of a waste not being able to share it with anyone.

It was a coincidence because as I finished that thought, some people did begin to show up on deck. I was glad to see this as I couldn't imagine being in a place where you could so easily see the comet and not take advantage of it. One of the people who came on deck was a professor of Astronomy at a college in New York whom I met earlier in the day. He was lecturing on board the ship about the comet, and I managed to catch the last 10 minutes of his talk. Thinking I was some amateur on deck, he asked me if I knew where to look to find the comet. Little did he know I knew what I was doing, so I pointed it out to him. I would have liked to see the expression on his face when I did, but it was too dark. After a few minutes, some more people came up on deck. I was glad to see that they were just later risers and did care to look at the great Comet Halley. It would be such a waste to travel to a place where it could be so easily seen and not take advantage of seeing it. There will be so many people who will miss seeing this once-in-a-lifetime sight because of their location in the world, the inconvenient time at which they can see it (such as the early hours of the morning), or just the lack of desire to see it. I didn't want to be one of them.

I couldn't believe that I could see the comet so easily. I was worried I would need binoculars or risk bringing my scope up here to see it. I couldn't have asked for a better night for my viewing. Everything was so brilliant, not even a moon (which was well below the horizon) to cast light that would have faded the comet's now mighty tail. One of the reasons I took this particular cruise was because the moon would have set by midnight. I couldn't have asked for more: the moon in the evening in the Caribbean and Halley's Comet in the early morning.

I wondered if this would be my best night to see the comet. My research showed that this is the furthest South the ship will go without being bothered by city lights, and as far as I could roughly figure, my present latitude was 12.21° N and longitude was 69.00° W, give or take a degree. (I figured this by knowing which port we were heading to and looking at a map) the ship's first stop is at the Island of Curacao, part of the Netherlands Antilles.

As difficult as it was to leave the best view I had of the comet to date. I decided to go back to my cabin and get some rest. For the time, I had my fill of the comet; for almost 40 whole minutes, I had it all to myself, and as it was nearing 5:00 am, the sun would be coming up soon to wash away the comet. I had seen what I came to see, and I was not disappointed; in fact, I was thrilled. As I was leaving the deck, a few more people came up to the observation deck, and as I was departing, I heard the first oohing and aahing when they first saw the comet. It reminded me of the 4th of July when people would watch fireworks; it was a pity that this particular firework came by only every 76 years.

Chapter 11

March 18, 1986

**Location: Caribbean Sea North of Caracas, Venezuela
Latitude +11.21 by Longitude -69.00 (roughly)**

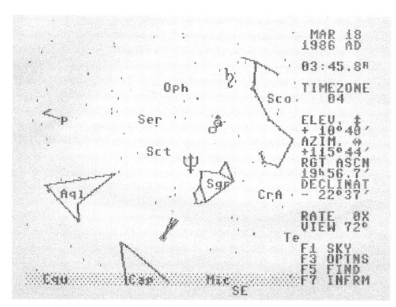

Scan of original Sky Travel star chart. Printed in 1986.

By 3:34 am, I was back on the observation deck. I wanted to get there a little earlier to see the comet as it climbed higher in the sky. I was highly shocked to see that there were people on deck when I got there. Maybe I shouldn't have been surprised. There was talk on the ship about the March 17th viewing and how easy it was to spot the comet, so maybe people on board were beginning to take more interest. Many people who asked me (including the friends I made on board the ship at my dinner table) were surprised to hear me say that you could see the comet so well. The word traveled quickly throughout the ship, and even those not on board specifically to see the comet were becoming interested. Comet fever was spreading.

From the time I arrived until I left the observation deck this morning, the number of people grew from twenty to fifty, not counting the people who came and went. Some people had the foresight to bring binoculars with them. They were kind enough to share with others, and I did ask to borrow them to look at the comet. It was another clear sky, and the comet was found easily hanging in the sky. It was still terribly windy here, and I took shelter behind a solid railing to protect me from the wind. It was there, out of the breeze, with a pair of borrowed binoculars, that I saw my first magnified view of Halley's Comet. It was such a wondrous sight. I would have had a much better view of it with my telescope, but the constant roll and pitching of the ship would still make it impossible to keep it in the field of view for any real length of time. Looking at the sky through binoculars was a different feeling. I was so used to using a telescope and one-eye viewing that it took a little getting used to, and I was still sorry that I would not be able to use my telescope.

In the distance to the South, I could see the lights of Caracas, Venezuela. It was our next port of call, and this would be the furthest South I have ever traveled. (But that would change in following decade). While still north of the equator, this is my best location to view the comet. Through the binoculars, I looked to see if the comet was different in any way. I noticed that the tail didn't seem as long, concluding that being this close to land and the lights of the big city affected viewing, even with binoculars. I was so glad I had seen the comet the previous morning; even without binoculars, it seemed more beautiful. The other people on deck were never the less impressed with what they saw, and I did not mention the better clarity of yesterday's viewing to anyone. The Astronomy professor was also back on deck, pointing out different constellations such as the Southern Cross and the Teapot (The constellation Sagittarius) and naming certain stars in the constellation Centaurus. Some of the other passengers were really into it; they, too, had never seen the sky so full of stars and couldn't believe there was so much to see. Depending on where you live, this could be a real viewing difference. When we went to the stern of the ship, where more deck lights were on, all you could see were the primary or brighter stars and nothing of the comet's tail. It's funny how walking 600 feet changed the entire view of the night sky.

I was glad that I saw the comet again and even more pleased that I saw it yesterday. I had it all to myself, then for a while undisturbed, and I could just let it soak in. This morning, it was also at this time that a crew member was passing around a paper for those on deck to sign with their name and cabin number. It would enable one to get a certificate endorsed by the cruise line saying that you had seen Halley's Comet on this particular cruise. It was a very generic certificate stating the week you saw Halley's Comet, not a time, date, or location. Even though I kept a record, getting something more official would have been excellent.

One last thing to mention on this early morning is that people had learned much about light pollution. If you were to walk up to the observation deck with a flashlight or a light of any kind or even turn on a light while on the observation deck, you took the chance of getting thrown overboard. Knowing that looking into a white light would mean you would need at least another 10 minutes for your eyes to re-adjust to the darkness again, and with dawn approaching, time was precious for viewing. I left the observation deck before the comet began to fade. I couldn't bear to see it go. I never told anyone that there was a better viewing the previous day; however, the professor did bring it up to me later on in the day to see if I noticed it. While he mentioned it to me, he said he didn't tell anyone else so they would not be disappointed in missing a better view of the comet, which was the same reason I didn't.

I walked back to my cabin, tired but glad that I again saw the great Comet Halley. To me, the cruise was worth it to fulfill this 5-year-old passion of mine. I would sleep peacefully knowing that if I didn't see the comet again on this cruise, I did accomplish what I wanted to do by taking this cruise.

Chapter 12

March 19, 1986

Location: Eastern Caribbean Sea
Latitude +12.00 by Longitude -62.00 (roughly)

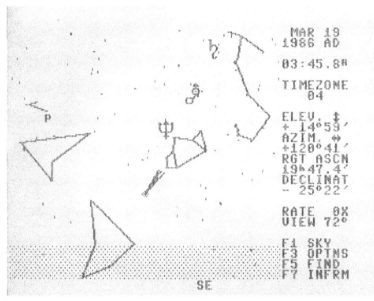

Scan of original Sky Travel star chart. Printed in 1986.

I went back to the observation deck at or very close to 3:45 am again. Walking wasn't easy because not only did I have the high winds and the roll and pitch of the ship to worry about, but also because I was carrying my telescope. There had to be between 75 and 90 people on deck this early morning. The sky was as clear as it could be, which made me glad that I would have another night of Caribbean viewing. Since the ship was traveling East-Northeast towards Grenada, we were far from the lights of Venezuela. Viewing the comet was probably just as clear as the first night I had seen it, but something inside me said it wasn't. The professor was again on deck, pointing out objects in the sky like the nights before. He knew I had my telescope on board from previous conversations we had. He told me, "You are awful gutsy to bring that up here in this wind." I replied, "What the heck? It is insured". I added, "I brought it all this way; the least I can do is try to look at the comet through it." And I did.

People were amazed to see the size of the telescope. Even I couldn't wait to glimpse the comet through it. It would be tricky, to say the least. Even without the wind, between the motion of the ship and the vibrations of the ship's engine, being able to focus would be a challenge. I would have to try to center the comet in my field of view, and each time the ship would pitch up, I would then focus and wait for the ship to dip and come back up. This system will work to a degree for viewing, but any photography would be out of the question. I was able to view the comet as the ship rode the waves; it was not ideal viewing, but it was something. Using my largest millimeter ocular and several tries, I focused on the comet. Halley was just a blur until I was able to focus on it, but it was there. It was the comet I had waited so long to see, and this was the biggest I would ever see. Even with the hardship I was dealing with. It was worth it; I could see the comet, its coma, its long-pronounced tail, so white and vivid. An astrophotographer's dream if you were on solid ground. But not for me, not here, not now.

Everyone wanted to look at the comet through my telescope. Some people on board knew what my telescope could do because I had it set up on a higher deck while we were in port in Caracas to look at the city and faraway mountains of South America. It was also my chance to ensure it wasn't damaged in transport. I hated to disappoint anyone, so I gave everyone who wanted to view it a quick explanation of how the mechanics of viewing would work with the ship's rolling. I would still have to re-center the scope every few minutes to keep the comet in the center field of view. I also let people look through the finder scope, which was attached to the main scope, which had a larger field of view. Here, the comet did not move out of the center as quickly; it was not as big, but it was larger than viewing it with the naked eye or smaller binoculars that others had. It was now I wished that I had a wider ocular with me. But I didn't have one. That would have allowed me to view it a little easier, and I would not have had to adjust my scope as often, but I wanted the greatest magnification of the comet that I could get. I still would not have been able to take photos, but I could have observed it better. It was a lesson learned all too late. Something else I learned a bit late is that you must take excellent care and protect your eyepieces as I dropped my largest ocular in the dark, and it rolled overboard. What a lesson I learned that night; at least I had the brains to insure all my equipment before I left the States, and fortunately, this was the only disaster of the trip.

I still enjoyed looking at the comet; to me, it never got old. I only had a few more days to view the comet from this Latitude and Longitude (which was 12° N Latitude by 63.50° W Longitude, as close as I could figure.) I would have a month, perhaps a bit more, after this cruise ends to see it back home, but viewing it would not be as nearly as spectacular as it was this week. My life might be a little different after the comet is out of view. This long-awaited obsession will have come and gone. I would like to know what I would look forward to doing after this.

There was more oohing and aahing from people coming on deck and seeing the comet for the first time. Some commented that it was brighter now than yesterday when they saw it. Being where we were did make a difference as the comet was in the sky over water instead of land with city lights below it as it was when we headed for Caracas. It was apparent that people on board were spreading the word about the comet as each night, there was an increase of star gazers that ventured out to look at the comet and the night sky. I was glad to see more people taking advantage of this. I wondered how many of the estimated number of passengers (which was right around 700) were going to go this whole trip and not take advantage of seeing the comet.

This morning, the comet was just as beautiful as ever. Still holding its own it the Southeast sky, still pointing towards the constellation Capricornus, its tail still reaching toward Sagittarius. Back in Wisconsin, Halley would be out of sight below the horizon. Everyone back there wouldn't be able to see this spectacular sight; even if it were high enough in the sky, the light pollution of the cities would greatly diminish the view of its tail. I would have really liked to have been able to take a photo to bring back to them, but it still wouldn't be the same. Though a picture may be worth a thousand words, this view was priceless to me, and there aren't enough words to describe what I was seeing and how I felt about it.

Chapter 13

March 20, 1986

**Location: Eastern Caribbean Sea
Between the Islands of Martinique & St. Lucia
Latitude +14.40 by Longitude -61.05
(As provided by ship's navigator)**

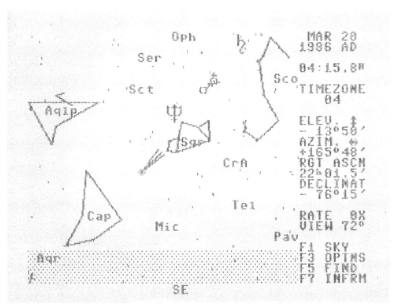

Scan of original Sky Travel star chart. Printed in 1986.

It was 4:10 am when I made it out to the observation deck, a bit later than usual. There were fewer people on deck than the night before, but still a decent turnout. A few more people had binoculars with them this early morning. The wind was stronger and appeared chillier than the previous mornings if you didn't have a jacket on. The reason I made it on deck later today was because I was talking with friends I made at the on-board casino. After the casino closed at 2 am, we sat around talking about our respective countries. I was from the US, and they were from the UK. My watch alarm went off at 3:40 am, and I explained to them it was time to go up on deck and see Halley's Comet and invited them to come along to see it. One of the girls in the group quickly and vocally declined, saying it was bad luck to see the comet. I myself had never heard that story or anything of that sort and asked her where she had heard it. She replied, "Where I come from, Sheffield, England, it was always said that bad luck or an awful thing will happen to anyone who looks at the comet." She asked me if I had seen the comet, and I told her I had several times. All she said was, "You better be careful then."

I asked her if she had ever looked at the night sky while on board the ship. I explained why a ship at sea was ideal for looking (certainly not photographing) at the night sky. The simple reason was that there was no light pollution or obstacles such as trees on land in the cities. It was hard to believe that she didn't or others from this group didn't take more interest in the sky. But then again, some aren't as interested in it, and they were here to work, not to vacation like I was. I had met some of my other friends from the casino the night before on deck before the comet came up, and I pointed out some of the better-known constellations to them, actually only the obvious ones that I knew. I still had a lot to learn about the night sky, and it was more challenging to pick some out because of the greater abundance of stars in the sky here.

They were surprised to learn how much there was to see in the night sky, and they at least appeared to take some interest. Perhaps a little bit of what I told them about sunk in, and they may look up at the sky a little more often now, realizing the advantage they have working on a cruise ship, giving them ample opportunity to see a sky devoid of light pollution. They did admit they never took advantage of this, at least not before I pointed it out to them. I thought I would have jumped at every chance I could, but then again, I was not working on the ship; I was just a passenger craving to see Halley's Comet. But it should be known, at least from what she admitted to me, and I believe her, that she saw the comet as much as she wanted to see it, and that was not at all. Some people like me traveled thousands of miles to get this opportunity for a few days, and she had it nightly and never took advantage of it. But enough of this comet tale.

The ship was traveling north to Martinique, and Grenada was now South of us. And the comet was behind the ship over the ship's bridge in the Southeast sky. It was still as beautiful as ever. I borrowed someone's binoculars for a moment to gaze more closely at this beautiful nighttime spectacle. I don't know what power the binoculars were; for some reason, I never bothered to ask. In the back of my mind, I couldn't help but wonder why the comet would bring bad luck. The only time I ever heard of a comet bringing bad luck was the one that hit the Earth and killed all the dinosaurs some 50 million years ago. If indeed, that is what happened. There were still a couple of theories on what wiped them out. Halley's Comet wasn't a threat; I knew it wouldn't hit the Earth.

Next month, when the Comet makes its closest approach to the Earth (April 11, 1986), it will still be 39 million miles away. I will be on the wrong side of the world to see it when it does, but it made me wonder how, in 1910, when Halley flew by the Earth at just under 14 million miles at its closest distance to Earth, it had to be such a marvelous sight. There was so much less light pollution than today and so much closer. It must have been a grand sight. No wonder people back then were so taken in by its vastness, and others feared it was the end of the world. Not the comet's next visit in 2061, where its closest distance to the Earth will be 44 million miles, but its visit after that in 2137, the great Comet Halley will come within 6 million miles of the Earth. (so I read). Imagine that, almost close enough to reach out and touch it. Hopefully, the light pollution won't be as bad, and viewing can still be as spectacular as it is tonight. Perhaps a descendant of mine may see the comet's appearance in the years to come. If one does, I can only hope they would have read this and learned that some ancestor of theirs had gone to great extents to view this once-in-a-lifetime attraction.

Getting back to the Caribbean skies, I observed that the comet hadn't changed much in appearance, at least from what I could tell. If I was on level ground and had my telescope, I could tell that the comet was moving by measuring the distance between stars. I'm sure my *Sky Travel* program could tell me in a minute. But I didn't have that with me; all I had was a single chart for each day of the cruise that I printed out and brought with me, and I never took those out on deck with me. It was too dark to read them, and they were only for reference before I went out on deck. The professor was back, pointing out stars and constellations as he did on past mornings, answering questions for those who didn't know what they were looking for or at. I heard people talking about a big telescope that someone brought to the observation deck yesterday and were hoping to see the comet again through it. I briefly thought about getting it, but it would be too much trouble and too late to make it worthwhile. The word did get around about my letting people look through it yesterday, as the ship's captain sent me a bottle of wine to my dinner table in thanks for allowing others to look through it.

I tried to look for the planet Saturn, but it just looked more like a bright star. Binoculars weren't good enough to see the rings while on a moving ship. Tonight, I also took time to look more closely at the rest of the sky, taking advantage of my present situation. There were other constellations that I was able to find, such as Hercules and Lyra. They were two of my other favorite groups of stars to look at. They both had interesting things to look at (not that I would see them here), but there was a globular cluster in Hercules and the Ring Nebula in Lyra. I had seen both of these with my telescope back home. The stars that made up their frames were much brighter than when I would look at them back in my city. This would be an ideal place to watch the skies. It's too bad there isn't solid ground under me here.

Something else I neglected to mention in previous entries is that you can see a white band of stars crossing the sky. It stood out like an endless white line against a black asphalt road. Not even when traveling to Northern Wisconsin have I seen a sky as clear as this. I only have two more nights of viewing the skies like this, and I felt fortunate because there have been no cloudy nights to speak of, with the exception of that first night. I couldn't have asked for more clear skies than this past week.

I felt at ease and relaxed as I stood on that windy observation deck. Though I came on this cruise for several reasons, I have found peace of mind and solitude in the skies above.

Chapter 14

March 21, 1986

**Location: Eastern Caribbean Sea, Southeast of St. Thomas
Latitude +18.21 by Longitude -64.55
(As provided by the ship's navigator)**

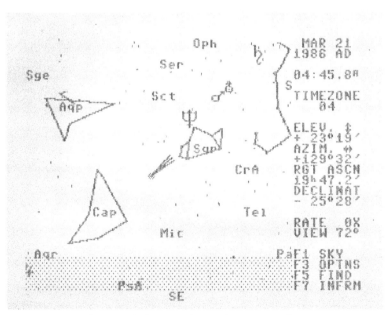

Scan of original Sky Travel star chart. Printed in 1986.

It was sometime yesterday afternoon (March 20) that I read on the ship's itinerary that tomorrow morning would be an exceptional morning to see the comet. The ship would be away from any island or port, like my first night at sea (Why this was never mentioned by the cruise line before, I don't know), true darkness with no lights anywhere except those on board.

Throughout the rest of the evening on board the ship, the word spread that tomorrow would be the best night of the cruise to see the comet. I ran across the professor and asked him what would make the next morning's viewing better than the previous days. He informed me that he was a bit disappointed with the few people taking the time to see the comet. He thought there would be a better turnout if he made it sound more spectacular. I thought about it for a while, and even though it was a little dishonest, it did make sense. What I mean is that those who had yet to take advantage of seeing and appreciating the comet wouldn't know if its view was better or worse than previous nights. Those who have already seen and enjoyed the comet will understand the reason for getting as many people out to see it as possible. Whether it was morally right or not, it most surely worked.

When I was halfway up the outer stairway to the observation deck, I could hear people talking and more people coming up behind me. I went up to the observation deck with a bottle of wine I bought at the ship's bar right before closing at 3 am. One of the guests from my dinner table brought some glasses. There were roughly 60 people on the deck by 4 am. when I got there and, more were coming. Some of them looked at the comet and left; others looked at it and stayed and enjoyed the night sky as they had never seen it before. One could tell who had not been to the observation deck before that night because, like me, they couldn't believe how beautiful the sky was. You could hear people talking about how big and beautiful the comet was, how they didn't know that there were so many stars in the sky, how impressive the sky looked, and how they had missed this opportunity earlier in the cruise. The professor's idea worked; never before have I seen so many people on this deck.

I opened the wine, poured some into a glass my friend brought, and made a toast to Halley's Comet. I said, "Probably not again in my lifetime, but for others in the years to come." I admit it wasn't much of a toast, but it seemed appropriate at the time. I guess you had to be there. Some lady said to us, "I see you brought something to stay warm with." I offered her my glass, and she accepted it and drank. As she gave me back the glass (and to this day, I can't tell what possessed me to say it), I couldn't help but say,

"That bottle of wine was bought by an old relative of mine when Halley's Comet came by in 1910). She couldn't believe that I had shared it with her. Some other people standing nearby couldn't believe I had brought a very old bottle on deck. It was good that it was so dark, and you couldn't see the bottle or read the label and learn that it was only three years old. Anyway, my friends and I passed around the glasses and kept refilling them. Everyone seemed to buy the story. One person commented that they thought it was great that someone had the foresight to do something like that. One of my dinner companions, who knew that the bottle was not that old, leaned over and whispered to me so that no one else could hear, "You do realize that these people are going to go back and pass on this story, thinking that they are drinking 76-year-old wine". I replied, "I know, and maybe I will feel a little guilty about it tomorrow, but they are going to remember this night, what they did, what they saw, and what they drank, and who knows, maybe they may do it to pass on to one of their family members."

Everyone on deck in those predawn hours was having a good time. It seemed more alive or vibrant than the other mornings. Another person said too bad the bars are closed; otherwise, we could have a real comet party, but I already thought we were having one. I thought you only had to be under the comet to have a comet party. I was with a group of people pointing out the constellations I knew, showing others where Jupiter and Saturn were, and making sure to point out the Southern Cross. The comet was in the same place in the sky as it had been the night before but in a different place from the ship. You had to go to the observation deck's far starboard(right) side to see it. The ship was traveling Northeast at this time. I could tell this by looking at the stars and checking one of the ship's compasses on deck.

The comet was there in the sky, now moving at over 100,000 miles an hour and increasing as it picked up speed in its journey around the sun. To think, I only had one more morning to view the comet after today (at least while in the Caribbean). I had a big decision to make right now; it was almost 4:45 am, and there was an extremely light tint of blue coming out of the horizon to the East. I had to decide if I wanted to see the sun come up and wash out the comet. I had told myself earlier in the cruise that I would wait until the last day of viewing to see the comet washed out by the sun's blinding light. It meant I would have to take a chance on the clear sky one more night. Up to now, I have been extremely fortunate; all but the first night had been perfect for viewing. Would I have one more chance, or would my luck run out? I decided to wait. Taking one more last look at the comet and the magnificent tail it was sporting in case this was my last chance to see it; I left the deck knowing that if I didn't see it again (like this in all its glory and without the light pollution I will have to deal with back home) I was satisfied with what I saw, for me taking the cruise to see the comet was the right thing to do.

Chapter 15

March 22, 1986

Location: Eastern Caribbean Sea North East of San Juan, Puerto Rico Latitude +19.00 by Longitude -65.00 (roughly)

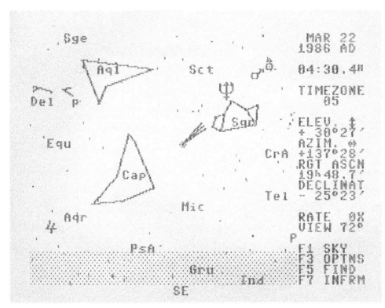

Scan of original Sky Travel star chart. Printed in 1986.

Today would be the last morning I would be able to see the comet from the Caribbean. I had been checking the sky periodically through the night, and it remained clear in all directions that I could see. When 4 am rolled around, I went to the observation deck for the last time. There weren't as many people tonight; as it was the last night of the cruise and all; everyone was pretty well wiped out with the late nights or early mornings. I know that I lost a lot of sleep in the past week, but I was running on adrenaline and knew I could muster through one more long day. I would sleep on the plane on the way home and catch up on rest after returning home to Wis. The late nights and early mornings were worth it to me in my quest to see the great comet.

The Comet Halley was once (and for the last time for me here) hanging so magnificently in the sky as it had all those previous nights (or early mornings) on board. The comet's tail was still opaque near the head of the comet as it was the first night of viewing, though it appeared to fan out a bit more. It could have been my imagination, but I didn't think so, and with no photos of it to compare it to, I couldn't check to see how different it did look. The comet has to be closer to the Earth at this point, now traveling at over 120,000 miles per hour after rounding the sun. It would stand to think that the comet would be larger than it was when I first saw it a week ago. Now less than a month away from its closest approach at 39 million miles to the Earth on this particular journey.

The rest of the sky was just as heavenly, so brilliant and full of stars in all directions. It reminded me of the ending of an old sci-fi movie made back in the 50s: The Incredible Shrinking Man. Near the end of the movie, before he shrank into the infinitesimal and ventured outside to the open world, he looked up into the night sky and said, "I looked up as if I somehow could grasp the heavens, the universe, worlds beyond number, Gods silver tapestry spread across the night." It fit as beautifully as the dark sky was, and this is the last time I would see it like this.

Like a camera, I looked as hard as I could with the naked eye; I tried to burn an image of the comet into my head, knowing that it would eventually fade or distort as most older memories do. I tried to stare at the comet long enough to last a lifetime as this particular view would have to last me forever. I didn't get the picture I had hoped for on this trip, and though a picture is worth a thousand words, this sight was indescribable.

Very soon now, the sun would begin to rise in the East, slowly but ever so surely, washing out the tail of the comet and then the comet itself. I took a look around at the rest of the sky. I stared again at the Southern Cross; this, for me, was still a sight to behold. I would have to return to a place where I could once again view it with ease. Like the comet, to capture the beauty that it holds and that so few understand. I looked at the constellation Sagittarius, the comet's tail reaching out to it. I suddenly noticed a slight glow on the horizon; the sun was approaching, and the night was beginning to fade into the dawn. I looked back at the comet while I could still see this heavenly spectacle and was grateful the sky was as clear as could be. I never took the opportunity before to watch the comet fade away; I would do it now because I wanted to see it as long as I could. Maybe it would've been better to leave the observation deck while it was still dark so I could have remembered the comet as it was, in all its magnificence, its tail stretching across the sky, millions of miles long and not watch it fade away by the sun's rays. There were very few people on deck now, some couples and a few people like me, all alone. No one was talking, just looking. They probably had things on their minds or were contemplating the last minutes of the comet or the end of the cruise and wanted to do it alone. I was surprised the professor wasn't here to see it one last time. As for myself, I just wanted to watch the comet. As its tail slowly started to shorten with the rising sun, I reminisced about the last week on board the ship, the things I had done, the day tours I took, the people I met, all the good times that happened, enough memories to last a good long time. It seemed fitting that the comet was fading with the end of the cruise. After all, I took this cruise mainly to see the comet. I admit that I silently cursed the fact that I lived in the Northern Hemisphere. While I was fortunate to be here during this particular pass of Halley's Comet where I was able to see it before and after it rounded the sun, I was living in the wrong part of the world to be able to view when it would be high in the sky during its closest approach to the Earth. Back in Wisconsin, its mighty tail will be erased from my view, even though it would still be there, it will be invisible to me and it will be so low in the sky once I return home. This will be last optimal viewing day and I had to accept that.

The sun quickly breached the horizon, and within minutes, all you could see was a bright dot in the sky where the comet was. It was tailless, and all the faint stars to the East where the sun was rising had already disappeared for the day. Soon, they would be followed by the comet and the brighter stars. In less than 10 minutes, all but the brightest stars, the comet, and the planet Saturn were left. The sky was pale blue, and the comet was nothing more than a speck. This is my second time seeing the comet in the daytime sky. It took just a few more minutes before the comet was completely gone from sight. I was sad to see it go; I felt a sense of loss, not unlike the passing of a dear friend, but not as traumatic. To think, I had traveled all this way to see such a wonderful sight, and now it was over. I was the last to leave the observation deck this morning, but as I did, I looked toward the sky in the area where I last saw the comet and, under my breath, said, "Goodbye, Halley." But what I should have said was so long. It wouldn't be goodbye, not yet. Maybe from this part of the world and under these unique circumstances, but I was still planning on seeing the comet again after I get back home. It would not be as majestic looking as it was here in the middle of the Caribbean, but it was still Halley's Comet and I was not finished viewing it yet.

Chapter 16

March 28, 1986

Location: Racine, Wisconsin. The Beach of Lake Michigan East of Latitude +42.46 by Longitude -87.46 (roughly)

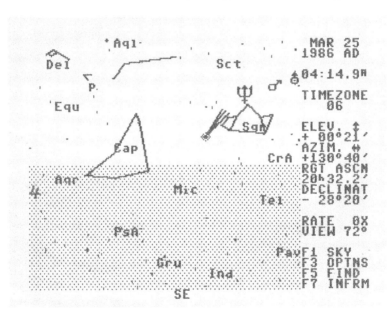

Scan of original Sky Travel star chart. Printed in 1986.

It has been six days since I last saw the comet. Seeing it again in Northern latitudes was indeed a disappointment compared to my last viewings. The fact that it was to be lower in the sky is nothing compared to the amount of light pollution in the area. The whole affair of seeing the comet was different now. The cruise was over, and there were no more late-night parties, crystal-clear skies, or walks on decks with the sight of the Milky Way Galaxy above. I set my alarm to a rude 2:30 am awakening. I called my friend Tom Cramer to accompany me on this damp, chilly morning. He rushed over, and we took my car to Chuck and Joan's house. They were friends of mine who live right on Lake Michigan. Some weeks earlier, Joan expressed her interest in the comet. She also had a friend, Judy, who cared greatly about the comet and wanted to see it. As the comet approached its closest distance to the Earth, comet fever was still alive and well. During this particular trip through our solar system, the Comet Halley will be closest to our Earth on April 11, 1986, and will be 39 million miles away. That is three times further than the 1910 appearance when its closest distance to the Earth was 13.9 million miles. It was said to be so close to the Earth that the Earth would pass through its tail. It was, again, bringing on a frenzy from those who thought it would hit the Earth. A popular scam back in 1910 was the sale of comet pills to protect you from the comet's dust as the Earth passed through its tail.

Chuck and Joan lived on the Western shore of Lake Michigan, only a couple miles from my house, making it a good area for viewing as the closest city lights to the East were from the state of Michigan, over 90 miles away. Still, I did have the light pollution from Chicago to contend with. The comet would be in the Southeast sky, and looking over Lake Michigan, which would be my best viewing area.

I brought my telescope to see the comet close up again. I didn't expect to see a sight like I did last week in the Caribbean. I was hoping to see the comet's tail, but trying not to get my hopes up. I wasn't counting on it, even though my *Sky Travel* program showed a very long tail on the comet. With Tom's help, I set up the telescope. One of the nice things about this telescope is that it is portable, even if it is large. Joan was already outside looking in the sky with a pair of binoculars, and just as we finished setting up the scope, Judy (one of Joan's friends) and her two daughters arrived to see the comet. The sky was clear as could be, and the comet was hanging in the Southeast sky where it had been for the past few weeks. I was several thousand miles farther North than I was last week, and the comet was much lower in the sky now. Using Saturn and Mars as reference points, finding the comet was easy. According to my *Sky Travel* chart that I printed out, Halley was still just below Sagittarius, with a tail stretching into the constellation itself.

I first saw the comet through my finder scope. It reminded me of the last time I saw the comet from Racine, but it did seem brighter now that it rounded the sun and was traveling faster and closer to the Earth. Using my most powerful ocular, I quickly focused on the comet. I desperately hoped to see the tail behind the comet again. The moon was high in the sky to the South, and its reflected light cast a glow all the way through the path of the comet. As I viewed Halley through my telescope, the comet was nothing more than a fuzzy cotton ball again, maybe a bit more elongated. Nowhere near the spectacular sight of a week ago, with no illustrious tail, just a fuzzy cotton ball. Needless to say, it was disappointing to me. That could be a poor thing to say; it's just that it lost its vigor from the last time I saw it.

I showed the comet to everyone there, and they all described it the same way as a wide cotton ball. They were amazed at what they saw; I kept quiet about the different views I experienced last week as to not make anyone feel as if they were missing out. They were thrilled when they first saw it, perhaps feeling the same way I did when I first saw it five months back. I pointed out some of the other celestial objects in the sky. I showed them the planet Mars and the rings of Saturn. A close-up of the moon really impressed them when viewing the details of the many craters it had, yet they were still thrilled to see the comet. Tom made some jokes referencing the movie *2001: A Space Odyssey* (his favorite movie) while looking at the moon, something about seeing a strange monolith standing in one of the craters. The others were thrilled just to see the comet, but did appreciate all the other sites I showed them. Seeing Halley was the focal point of today's viewing for them.

I remembered the first time I saw it; so, I think I knew what they felt. Deep down I wished they could have seen it like I did while in the Caribbean. Taking the scope to the beach that morning was worth the trouble. I showed the great Comet Halley to 5 more people, not just as a point of light in the sky, but up close. A historic view that they would remember.

Chapter 17

March 29, 1986

Location: Racine, Wisconsin. The Beach of Lake Michigan East of Latitude +42.46 by Longitude 87.46

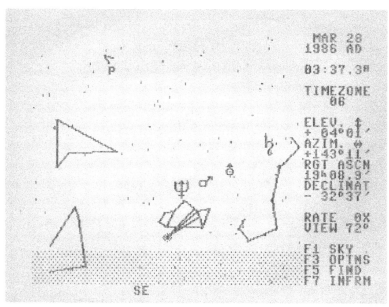

Scan of original Sky Travel star chart. Printed in 1986.

It was an early Saturday morning; I had set my alarm to wake me up at 2:45am. Plenty of time to set up the scope down by the beach and once again see the comet. I admit I was getting tired of these early morning viewings and losing a lot of sleep. But the clock was still running on how many more days I will be able to see it. Strike while the iron is hot, I thought. Once the comet is out of viewing range there is nothing I would be able to do about it. I must take advantage of all the clear nights I can and make sure that I show the comet to all that want to see it. As the comet picks up speed and reaches its closest point, it will be gone out of sight in just over a month from now or less. The Earth will be traveling in one direction and the comet the other, leaving much quicker than when it arrived some 5 months ago. It was now traveling at a speed of over 120,000 miles per hour, more than doubling its speed from the 48,000 plus miles per hour it was traveling at when I first saw it.

This morning, I would not be alone. I picked up my older sister Sue and my oldest nephew Eddie. After this morning all my nephews would have seen the comet. I would be able to give them the patches I sent away for. They said Halley's Comet 1985-1986, one of the few comet souvenirs that were being circulated during the days of the comet. I found the order form in the box containing my *Halley Project* game. By 3:20 am, I was back at the beach behind Joan's house, they were away on vacation, but gave me permission to star gaze at their house anytime.

The moon was not quite full, but it was as bright as could be in the Southern sky and the glow from the distant lights of Chicago were still visible to the South, even with the bright reflective light of the moon. Finding the comet was just as easy as the day before. It was still very low in the Southeastern sky with the moon not too far off to the right of it. I didn't expect to see its tail with all the light in the sky. I centered the comet in the middle of telescope and turned it over to my nephew. With great anticipation Eddie looked through the telescope. He was so excited and happy to see it and perhaps the most thrilled of all my nephews to view it. He described it like everyone else, as a fuzzy cotton ball. I wondered if Eddie would be able to see the comet again on its next pass in 75 years. He would be 87 years old then and his chances of a second pass would be still much better than mine.

My sister Sue looked next at the comet and asked where is the tail. I should have given them warning on what to expect. I think she was disappointed that she could not see the tail as it was so prominent in the photos in the paper. I couldn't blame her, after all the tail is the most impressive part of the comet to see, at least to me. After seeing the comets tail for a whole week and then viewing it without it were two completely different sights, yet deep down inside, I was still very much impressed. I re-centered the telescope several times so they could look at it as long as they wanted and then to make this early morning venture a bit more worthwhile. I showed Sue and Eddie the planet Saturn, as the rings of Saturn have never failed to impress me and I think Sue and Eddie enjoyed looking at Saturn more than looking at Halley's Comet. After all, the rings of Saturn are more impressive than a fuzzy cotton ball, wouldn't you say?

We weren't out viewing terribly long. Maybe just over 35 minutes. Eddie was still a bit tired, I didn't blame him, both he and Sue thought that it was worth it, to get up early and stand on chilly beach to view something that basically occurred once in a lifetime. But before leaving, I asked if I could try taking a couple of pictures of the comet as this could be my best last chance to do so. It didn't take me long to hook up the camera and with no electricity to power the clock drive, any time exposures would have to be just a few seconds long at best. I won't know how the photos would turn out until I get the film developed.

In another 2 weeks, Halley will make its closest approach to the Earth. It will be so low in the sky in my part of the world that I may not be able to see it, but I have come to the conclusion that after today, my best days of my viewing have now come and gone.

Chapter 18

April 6, 1986

**Location: Between Racine and Kenosha, WI.
On the beach of Lake Michigan.
Southeast of Latitude +42.46 by Longitude -87.46**

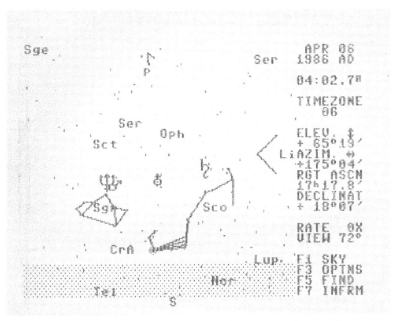

Scan of original Sky Travel star chart. Printed in 1986.

It had now been a week since I had last seen the comet; I picked up Joan's friend Judy at about 2:15 am. She had some friends who lived on the beach South of Racine and North of Kenosha, less than 10 miles away. I was hoping that this would be an ideal spot to see the comet because there would be less light pollution from the city of Racine, as I would be looking to the South; Kenosha was a little smaller of a city and would offer a little less light pollution not that I could really notice and I would be looking away from it. Unfortunately, the comet would be extremely low in the sky as it reached its closest approach to the Earth during this pass. As accurate as my *Sky Travel* program was, it showed the comet so low in the sky that I needed an unobstructed view, meaning no houses or trees. Being on the beach, I had a flat-level view in front of me. But to make matters worse, from where I was viewing it this early morning, it would put the comet very close or directly over the city of Chicago. The comets position was moving so much more now that it was closer to the Earth and because its speed was still increasing after the slingshot affect of it rounding the Sun. Because of this I could no longer just look into the sky and know where it was. There was some minor searching involved.

I quickly set up my telescope. I was very proficient at it since I had taken it apart and put it back together more times in the last six months than I had since I first bought it. I started searching for Halley's Comet by using Saturn and Mars (which didn't move as much) as positioning points and quickly found the comet. It was just West of the constellation Sagittarius. I remembered how it was on the East side of Sagittarius just a month ago. The comet looked like it did so many times before—a fuzzy cotton ball. The comet was surrounded by blackness, which I was surprised to see; there was even a touch of gray to the right of it, suggesting a hint of where its tail would be, opposite of the direction in which the sun would be rising. Perhaps because it was approaching its closest proximity to the Earth, but there was none to be seen, at least by me and my telescope. A timed photograph might say otherwise. I thought for sure with its location, it would be washed away with light pollution from the lights of Chicago, but I was wrong.

Judy and I took turns looking through the telescope with my various eyepieces, trying to make the comet as large as possible to view and always hoping to see its tail. I told her about what a beautiful sight it was during last month's Caribbean cruise. Again, I tried again to take some pictures of the comet, using more of the film in the camera that I was using last week. I still had to wait till they were developed to see if I had any positive results.

Clouds were coming in from the West, so viewing time was ending, whether I liked it or not. The temperature was still quite cool, which made the skies clear, and I was grateful for another night of seeing the comet. My days of viewing it were so very numbered now as the comet would be traveling below the horizon and out of my sight. By 4:15 am, clouds covered the sky, so I packed up the scope and took Judy home. As I was driving home from her house, I turned up the heat in the car and opened up the sunroof. The clouds in the sky were breaking up a little, and every now and then, you could see a star shining through the openings.

I wondered how many light years it took for the light of that particular star to reach this point. I asked myself how many years it would be before man could travel to a nearby star. I know it wouldn't be in my lifetime, but if I had a chance to go, would I? I don't know, probably not. It has been 14 years since man last walked on the moon, our closest celestial neighbor, and there are no plans to return to the moon. I may see man travel to the moon again in my lifetime. Could I see it with my telescope, I don't know, but I would give it a try?

Chapter 19

April 16, 1986

**Location: Between Racine and Kenosha, WI.
On the beach of Lake Michigan.
Southeast of Latitude +42.42 N Longitude -87.46 W**

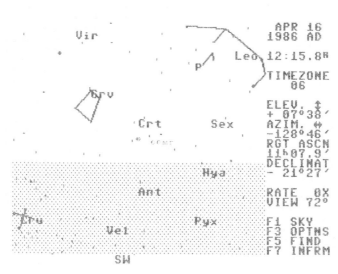

Scan of original Sky Travel star chart. Printed in 1986.

It was ten days ago that I last saw Halley's Comet, and it had been five days since it reached perihelion with the Earth. That is the closest distance that it would be to the Earth, and that would be 39 million miles. Since then, the Earth, which is moving in its regular orbit, would be traveling in the opposite direction of Halley, now traveling back into deep space past the distance of the giant planets of our solar system. Halley's Comet was now moving at over 122,000 miles per hour, the equivalence of covering the distance between the Earth and the moon every 2 hours, and it was getting smaller in viewing size with each passing day.

I would have liked to have seen the comet at its closest point to the Earth, but for me, here it was below the horizon. Even though my last viewing was five days before it reached perihelion, the comet would have been larger than seeing it now, five days after perihelion. Each day, I delayed seeing it; the smaller the comet would eventually become and out of the viewing range of my telescope. As beneficial and as good as my telescope is, it still had limitations, and I feared I would soon reach them. Once again, the clock was running, and now it was running faster than its arrival here, at over 122,000 miles per hour.

Joan's friend Judy, whom I met a couple of weeks ago, had friends that lived on Lake Michigan between Racine and Kenosha, making the sky a little darker with less light pollution and, more importantly, a flat, unobstructed view over Lake Michigan. She offered to let me take her there, and I jumped at the chance; it would be the best place to attempt to view the comet right now. I picked up Judy at her house at about 11:30 pm, but if things could go wrong tonight, they would, and they did. Because of the comet's increased speed, I had to check my *Sky Travel* program daily to know precisely where it was. According to my *Sky Travel* program, the direction of its tail had switched directions and was noticeably shorter as I looked at it on my computer screen. For today's viewing, I was a little careless when putting in my latitude and longitude on this day. While it might not seem like a big deal, it was because it showed the comet higher in the sky than it actually was, but I learned that after I had set up the telescope near the beach. Right away, when we reached the lake house, I noticed how heavy the air was with moisture, and there was fog over the lake to the South in the direction I would be viewing. The moon was half full in the opposite direction to where I would be looking; the only good thing going for me tonight.

I was looking for a familiar constellation to help me find Halley, but it was difficult with the air quality tonight, and when I did find a familiar sight, in this case, it was the planet Saturn; it was much lower in the sky than anticipated. Halley's Comet, I knew, was much lower than Saturn, and it was then I realized I would not see it tonight; it was either lost in the low levels of the fog or too close to the horizon for viewing, making tonight's venture a total loss. It took a few minutes for it to sink in that I would not see the comet, and just days past its closest point to the Earth. I was frustrated, to say the least, and I was also hoping to try to take some more photos of it. I never even bothered to take out my camera, which I later learned I left at home. So, even if the weather was favorable for viewing and photography, I didn't have the means to do it.

Because of the moisture in the air, the dew was rapidly coating my lenses. This is something that has happened in the past, but not this quickly and not as heavy. There was not even a wind tonight to blow the fog away. Yes, tonight was nothing short of a waste of time. It was then quickly decided to call it quits, pack up the telescope, and go home. The lens was completely and heavily coated in the falling dew, and I did not relish packing it up wet. I didn't even have a towel or anything to wipe it down with. I took Judy home, drove back to my house, and then once again carried my telescope inside, unpacked it, and wiped it down so it would rust or corrode.

It was nearing 1:30 am, and as tired as I was, I did not hurry to clean off my telescope. I didn't have to work in the morning, which was in my favor. Still, as I finished drying up my telescope and packed it back into its case to take out to wherever my next viewing would be, stark reality suddenly set it; my time with Halley's Comet was nearing the end. I waited four years for its arrival. I followed it in the newspapers, magazines, and on television for the last five months, and now, during this month of April in the year 1986, it was finally coming to an end, and it was just dawning on me. I might only have a couple weeks left to see it, maybe even just days. I tried not to think of what to do once the comet was gone; what would my next quest or venture be? What would I look forward to? I had no idea.

Chapter 20

April 22, 1986

Location: Modine-Benstead Observatory. Union Grove, Wisconsin
Latitude +42.43 by Longitude -88.01
(As provided by observatory staff)

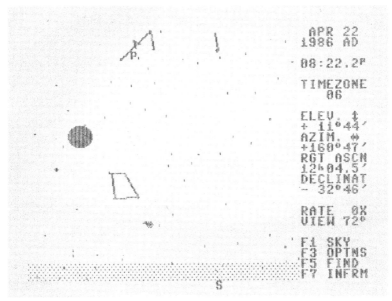

Scan of original Sky Travel star chart. Printed in 1986.

It has been two weeks, actually 16 days, since I last saw the comet; needless to say, many changes have occurred with it. Since I last saw it on April 6, 1986, Halley's Comet made its closest approach to the Earth on April 10-11, 1986. It was just below the horizon for me to see on those days and cloudy skies have prevented a couple other opportunities I had to see it, or at least try to see it. The newspapers and *Sky and Telescope* magazine said Australia was the place to be for the best viewing. Going deep into the outback, away from any city lights, would make the best viewing. As much as I would have liked to have seen it on this historic appearance, logistics and cost deeply prevented that from happening. Also, my work schedule for the fire department, along with cloudy skies, prevented me from being able to attempt to view it. I heard that some people from my hometown went to see it on the 11th, but it was cloudy; I felt so bad for them. My optimal time of viewing the comet has come and gone, and now I'm just trying to see it a few more times before it leaves me for good.

According to my ever-faithful *Sky Travel* program, the tail changed its direction from behind the comet to in front of it as the comet was now heading away from the sun and now traveling at over 120,000 miles per hour and the program that showed the tail to be so long was quickly shortening with each passing day. The comet was again a nighttime viewing object for me. No more getting up in the early hours of the morning to go and view it. I took my telescope out to the Modine-Benstead observatory, home of the Racine Astronomical Society. It was roughly 15 miles West of Racine, where they have a 16-inch Cassegrain telescope, twice the size of my telescope. The observatory was open to the public on various nights, more often during the days of the comet. It was usually open on the nights I was at work at the fire station, or else the weather was terrible, filled with cloudy skies, hindering my visiting there. This was my first and only trip here during the days of the comet, though I should have utilized this area more often. I also learned it was encouraged to have private citizens bring their own telescopes out here.

By 7:45 pm, my telescope was set up and ready for use. It didn't seem like this would be a good night for viewing because the comet was still low in the sky, and the moon was about as bright as possible. Adding insult to injury, there were clouds in the West moving in. Finding the comet tonight with the moon shining directly at me and into my telescope was extremely difficult. I had found it once, but I lost it with a slight foolish bump to the telescopes tripod. In my haste to find it took me almost 20 minutes to find it again. The clouds were moving in closer to the section of sky Halley was in, so now it was a race against the clock as well as Mother

Nature. The comet was 8 degrees off the South horizon, meaning it was not very high above the ground. It had been a challenge to view the comet as the light from the moon diffused its brightness. It seemed smaller in size to me, noticeably different than when I last saw it two weeks ago. It was true; Halley's Comet was leaving, and now at a speed of over 122,000 miles per hour. It was moving so much faster now since it made its trip around the sun, and as quickly as it came, it was leaving so much faster. What I had waited and prepared for the past few years and watched for the past 6 months was coming to an end.

The clouds miraculously stopped moving to the East. They seemed to hold a position about 15 degrees West of the comet, and I was grateful. Then it suddenly sunk in: tonight was probably the last time I would have to view Halley's Comet. It was heading away from Earth at a tremendous rate of speed, and it was constantly getting smaller with each passing day. I should also mention that the impact of viewing the comet was leaving as well. As much as it was publicized over the last six months, and since it had passed its closest approach to the Earth and could no longer be viewed with the naked eye as it was back in January, comet fever had cooled down and was no more. As I looked through the telescope, staring is a better description; comparing it to the first time, and many times, I looked at it, realizing it was coming to an end.

I reset the scope so that the comet was again in the center of my field of view. I stared at it as hard as I could, trying to remember everything I could about how it looked tonight. It was hard to see it now, especially after seeing it like I did in the Caribbean. I looked at my watch and noted the time; it was 8:22 pm. I looked through the eyepiece and at the comet for another minute as the Earth's rotation moved it out of the telescope's field of view. I then put the lens cover back on the scope, thinking this is it, the last time I will see Halley's Comet.

Chapter 21

April 28, 1986

Location: Racine, Wisconsin. 1024 Florence Avenue. Latitude +42.46 by Longitude -87.46

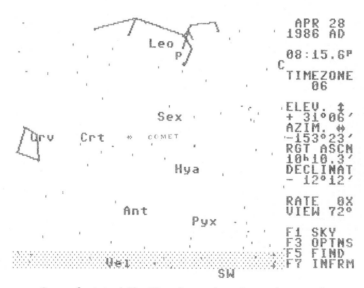

Scan of original Sky Travel star chart Printed in 1986.

It was six days ago that I last saw Halley's Comet. The media hardly mentioned it anymore. There was such a buildup in the months prior to its arrival and now less than 3 weeks since its closest approach to Earth, it was barely mentioned anymore. The comet was on its way out and, with it, the end of one of the only real long-range goals I had planned in my life, and it was leaving at over 122,000 miles per hour. Meaning that it could be about 60 million miles away by now. It was almost 8:00 pm, and I noticed the streetlight across from my house was out again. I thought I should take my telescope out and attempt to find the comet. In the city with the light pollution, its low angle in the sky would make it extremely difficult, if not impossible, to find, yet with nothing to lose, I decided to try one last time to view the great comet, Halley.

I first checked out the location of the comet with my *Sky Travel* program, which showed the comet's approximate location and showed it with no tail at all. After all, the tail would be pointing away from the sun, and the comet was a mere dot in the center of the chart. It still gave me a general idea of where to start my search, so once again I pulled out my telescope. Using my 18mm ocular, I slowly scanned the skies. I was looking due South, just below the constellation Alkes, "the Cup," and west of Corvus, "The Crow." Being ever so careful, I checked out each star, no matter how faint or small. There was no moon tonight as it was well below the horizon at this time, a significant advantage to my cause. Just as I thought I would never find the comet, I did. I noticed a very small, fuzzy glow near the center of my field of view. Like a cotton ball in a very dim light, it hung there in space; there was no tail on it, nor had I expected to see one. I gazed for a minute, pleased to see the comet a last time. It had significantly decreased in size over the previous six days, and it was purely a miracle that I found it at all; it was so much smaller in size than when I first found it last November—using my best oculars and Barlow lens (a lens for doubling the size of what you were looking at). It may be possible to see it again after tonight, but it would be much more difficult to find and time-consuming to do so. I knew deep down inside that I would not see the comet again after tonight, nor would I try to look for it.

It was from this very spot that I first found the Comet Halley 6 months ago, and it would be the last place I would see it from. I found that notion coincidental, a minor reference to Mark Twain's statement that he came in with the comet and would go out with it. I adjusted the telescope ever so gently to keep the comet in the center field of view. I l Looked around, none of my neighbors were out for me to at least ask if they wanted to see Halley's departure. I looked or maybe staired at it again realizing that this was it, my last look at Halley. I looked up into the sky to see what else I could look at and then thought against it: leave tonight for the comet and forget the rest. Tonight was the final end of my goal; the quest was officially over. I had one photo, not the best, but it was mine. It was taken on one of the nights in late March after my cruise while I was down at the beach in Racine. It was my own fault for not keeping better records of when each photo was taken. I had a large number of photos that didn't turn out as I experimented with time exposures, F-Stops, different ASAs, etc. I was fairly positive that it was taken on March 29, 1986, when I showed Halley to my sister Sue and nephew Eddie.

I had seen the comet in its different stages and from different locations, not just from the city of Racine but different parts of the Caribbean with clear and dark skies. I had seen it in ways many had not been able to or had not cared to. Because of my quest, I learned and saw about the beauties of space and saw one of its most spectacular recurring sights as it swung by the Earth every 76 years to say "Hello." Since the comets first recorded visit in 246 B.C. I was one who had the pleasure (and desire) to witness it. I was not one of those who felt it was taboo to look upon this once-in-a-lifetime event. It suddenly dawned on me that there were more people alive today than there was in 1910 meaning Halley's Comet could have been viewed by more people than ever before.

My Comet Tales

I put the lens cap back on the front of my telescope. Signaling that this time was my final viewing of the comet and my final goodbye to it. I put my telescope away knowing that it may be a while before I take it out again. I wanted to think that with maybe a lot of luck, I might be around in another 76 years to see it again, but who was I trying to kid? I'll be long gone before Halley's return in 2061. Perhaps one of my nephews will be able to see it, or I'll have children of my own so I can relate my comet tales to and possibly create a spark or desire to see it when it comes around again. Perhaps a copy of this journal will interest them, and they will be able to experience the joy and excitement that I had in my quest to see the comet. As I sat on my front porch looking up into the night sky, I reflected on the past weeks and months and of all the times I viewed the comet. The years I waited for it, the preparation and execution, and finally, the closure. I read somewhere that Halley's Comet passed through 6 constellations during the months of November through April. The last thing I have yet to do is make another entry in my journal of tonight's viewing. It really isn't a journal per se, but daily entries on my Commodore computer's word program. I'll print them out and keep these *Comet Tales* together. My quest completed and a goal now fulfilled and over, and I owe it all to Edmund Halley, the man who first determined the return of the comet and, unfortunately, was never able to see his prediction come true. He died in 1742; the comet's subsequent return was in 1758. I owed all this to him, a man from the past that I never met, who made a great contribution to the field of astronomy. The person who made this whole project possible for me, and I silently thanked him.

Chapter 22

In the Wake of the Comet

In the first few days after my last viewing of the comet, I considered taking my telescope back out to try to find it. I debated traveling further away from the city, possibly back to the observatory where there was less light pollution, in an attempt to see it. But why? I already said my goodbyes to the comet, so why do it all over again? It would only be a disappointment if I did not find it. My quest was over- bucket list item checked. I would still look in the sky in the direction I last saw the comet at night, knowing I wouldn't see anything, especially with the naked eye. Friends and co-workers still asked me about it for the next few weeks, but it was not in the news anymore, though an occasional picture did show up. Throughout the next few months, *Astronomy and Sky & Telescope* magazine would run stories about it or print someone else's photos of the comet they took. My photo did not compare to any I had seen in the news or other publications, but it was special because it was mine; I took it.

I pulled my telescope out less and less frequently after the days of seeing the comet, and somewhere along the line, I lost interest in the night sky. Maybe I burnt myself out a little with all the constant packing and unpacking of my telescope, moving it from all the different places over the past 6 months. I looked at the planets occasionally when they were traveling across the earlier hours of the evening sky and would still try searching out other deep-sky objects, but doing that from the city was difficult due to light pollution. Surprisingly, my desire to look into the night sky diminished, was lost, and remained lost for years. Since those days, I have married and moved to a new house near a big field where I could use my telescope whenever I wanted to. Unfortunately, I didn't want to look at the sky as much, and that was how it had been for a number of years.

In the late 1980s and early 90s, I spent seven years traveling around the world, climbing mountains for charity. I went to some very remote locations on Earth and once again saw the night sky without any light pollution. I was climbing some of the highest mountain peaks in the world, from Mt. Kilimanjaro in Africa to Mt. Aconcagua in the Andes of South America. My most remote climb away from civilization took me to Mt. Vinson in Antarctica. But I was there, during its summer where there was 24 hours of daylight, so I never saw a night sky there. However, I did see a night sky again from the previously mentioned peaks, along with Mt. McKinley in Alaska and Mt. Elbrus in Russia. During these climbs, I once again saw the dark skies dotted with the many stars I had not seen since March of 1986 when I took my Comet Cruise. It brought back memories of Halley's Comet that I shared with my fellow climbers. I wrote the book *Climbing for Causes: A Personal Story* using the journals I kept while on the mountains, in the same fashion, I kept pages on Halley's Comet, except on the mountains, they were all handwritten in blank booklets while laying in my sleeping bag at night.

Jumping ahead some 30-plus years, I have witnessed many things that have changed in the world since what I then called the days of the comet, including technological advancements. One of the most significant advancements that significantly hurt the backyard astronomer was the LED light, which uses less energy but is much brighter. The newer versions of the Celestron 8 telescope, while still having the same basic framework as mine, now has a much-improved clock drive, which allows the operator to more easily find, track planets and deep space objects for astrophotography. In the early 2000s, cameras changed from film to digital, providing immediate results of your work-there was no longer a need to buy and develop different types of film with various ASAs. While digital cameras were more expensive, film cost was also erased. I wonder what more technological advancements will be made by the time Halley makes its way back here in 2061.

During the Covid-19 quarantine days of 2020, with many people sheltering for those long months at home, I took out my telescope again.

It had been five or more years since I looked through it. I purchased an adapter for my digital camera and again tried taking photos of the heavens, but without a good clock drive, I could still take only short-timed exposed photos. What a marked improvement over my film days. Cameras explicitly designed for telescopes that you hook up to computers are now part of the norm for astrophotography. I don't think I will go out and buy one, but I have recently enjoyed taking some photos of the night sky. One thing I did learn is that in the past 25 years, since I saw Halley's Comet, light pollution has grown with the population. The areas of dark fields where I used to take my telescope when I first bought it had been developed with homes and businesses. The aforementioned LED lights were everywhere, including streetlights and porch lights, and you can now see a much stronger glow over the city at night. I can no longer see the horsehead nebula in the constellation Orion while looking through my telescope; I can only see it in longer-time exposed photos due to light pollution.

My *Sky Travel* program, which I owed so much to in my search for the comet, was outdated and phased out many years ago. I found a working version of it on the internet, checked the latitude and longitude settings I used for my local 1985 viewings, and was pleased to learn that I was less than 10 miles off when setting up my location. The latitude and longitude that I used on my *Sky Travel* program for my first sighting of the comet was +42.46, -87.46 when in actuality was really +42.77, -87.79. Newer laptop computers are now available, and other astronomy programs have replaced them, making it easier to look into the sky for planets and deep-sky objects such as nebulas and galaxies. With the internet now at your fingertips, and on a phone no less - who would have believed that back in 1985- you can pull up the program *Stellarium* (a modern version of *Sky Travel*) anywhere at any time, and it automatically tracks your location to see what is above and all around you in the night sky. I even get an email alert whenever the International Space Station will be flying over my house. But I'm still grateful for those non-technological, warm summer nights when a full moon shows itself with a few clouds in the foreground, making for a most beautiful evening sky to look at.

In July 2020, I learned about Comet Neowise and its prominence in the Northern Hemisphere. Sadly, I learned about it too late, and when I tried to find it with the help of my daughter, Rachel, we were unable to see it when viewing was optimal. She would come out to the field with me to watch various meteor showers over the years and even helped me look for other passing comets.

I don't know what made me do it; perhaps it was the announcement of Comet Green, but in January of 2023, I decided to take out my telescope once again and try to find it. The best viewing was between January and February of 2023, but due to its location over the city of Milwaukee, its light pollution washed it out; I could only see it in a short-time exposure photograph that I took of it. It was cloudy or snowing during its best days of viewing, preventing me from seeing it during its closest approach. It would pass by just under 26.4 million miles from the Earth.

While searching for something else in my office the following month, I found all I saved on Halley's Comet. I had old Newsweek magazines, newspaper articles, comics that I clipped about the coming of the comet, some commemorative Halley's Comet magazines, and even an old VHS tape hosted by William Shatner which gave some details about the comet as well as when and how to find it in the sky. Most importantly, I found my original notes and *Sky Travel* charts about my days viewing Halley's Comet, which I printed on my old Commodore computer in 1985 and 1986. I wondered why I never did anything with these notes or didn't remember what I had planned to do with them after writing them. Maybe because I couldn't find a basic book on Halley's Comet back when I needed it in 1986, and I thought I could write one, I don't know or remember for sure. Many books at that time were primarily technical or gave a brief history and charts on where to find it in the sky. I think I was looking for something more personal. I'm not sure this is it, but it was something.

I took the time to go over those old and now faded papers from the mid-80s, and after rereading them, I wondered where Halley's Comet was at that time. I then realized, with a bit of research, that on December 9, 2023, the great Comet Halley will be roughly at the halfway point of its journey from its 1986 appearance and starting its return to our Sun. Reaching its aphelion, or its farthest distance from the Sun, which would put Halley's Comet just over 3.25 billion miles away, slowly starting its 37-year journey back to the Sun for its next visit in July of 2061. Its speed again slowed down to just over 2000 miles per hour. For those around in 2061, the Earth will be on the same side of the Sun putting it in a much better position for viewing. (It will be a more impressive sight than the 1985-86 passing). Halley's viewing time will also be shorter than the 1985-86 passing, with viewing between mid-June through the end of August 2061. Its closest approach to the Earth will be on July 29, 2061, the day after it reaches perihelion (its closest distance to the Sun) on July 28, 2061. According to the internet on *Stellarium*, at least from the mid-North American continent, Halley's Comet will be visible in the early morning hours in the Eastern sky above Saturn, so it should be easy to find.

I will not be here for that, nor will my nephew Eddie, who unfortunately passed away, but my nephews Andrew and Benjamin, who were born at the right time, have a good chance of being repeat viewers some 27,500 days since their last viewing. My daughter Rachel will also have the opportunity to see Halley's Comet during this pass. She has the chance to be the second generation of my family to see it. While I doubt, she will have the same interest in it as me, and this may not make her bucket list, Rachel will still have the opportunity to see something that I was so thrilled to see in my younger days and which was very important to me then.

I also found an original version of The Halley Project- the game I had so much fun completing in 1985. With the help of a proper emulator, I was able to re-experience it on my present computer, and it brought back many memories of when I first played it. Once again, I completed its challenges and was again given a code number to send to Mindscape, the game's maker. But they, like Halley's Comet, are long gone.

In May of 2023, I should mention the Eta Aquarids, one of two meteor showers created by debris from Comet Halley. The Earth passes through Halley's path around the Sun again in October. This makes the Orionid meteor shower, which peaks around October 20th. So, if I saw a meteor during this time, would I be seeing part of Halley's Comet again some 37 years after I first viewed it?

In the past year of putting this all together, sifting through the magazines, articles, and other items I kept from Halley's 1985-86 visit- as well as reading the past entries of my comet tales- I was again reminded of something I had put aside over the past decades. The fact was there was still so much out there to see in the night sky. It may be harder to find them now with the changes in the world, but they are still there. I have been considering buying a newer Celestron telescope with a built-in clock drive to help me photograph some of the faraway galaxies and nebulas. I might even get a better photo of a passing comet than the one I took of Halley. It might be time to get out under those clear, crisp skies and take another long, hard look at what was and still is right above me.

While I'm no longer the child I was back in the 60s, I can still afford to have the same desires and thrills to find out what is in the night sky and search out some of the galaxies, globular clusters, and nebulas I found so many years ago. I think back to the final words of Jack Horkheimer's Star Gazer- "Keep Looking Up," I stopped doing this for a while but started to appreciate it again. With my daughter Rachel, I still enjoy going out and trying to catch some of the Perseid Meteors when they make their yearly appearance, and I still enjoy seeing the ISS as it passes overhead while it circles the Earth. Who knows, maybe I'll see and photograph another comet someday? However, I'm sure it will not be as exciting as it was seeing Halley's Comet years ago. The 1985-86 viewing of Halley's Comet was a once-in-a-lifetime dream (come true). But it would still be great to see another, and that would be another comet tale.

Chapter 23

Halley's Comet Facts

Halley's Comet, whose official name is 1P/Halley, was named after English astronomer Edmond Halley, who predicted it return in 1758. The P stands for periodic proving that comets circle the sun. Unfortunately, Edmond (or Edmund) Halley passed away in 1742, 14 years prior to its return, hence giving him the honor of having the comet named after him. Halley's Comet is the only short-period comet that can be visible to the naked eye from the Earth, appearing every 74–79 years with the average about every 76 years. The change in years is due to the gravitation pulling and pushing of some of the larger planets such as Jupiter.

Halley's Comet is roughly 9.3 miles in length and 5 miles wide in size. It is said to be made up of ice, dust and frozen methane. As Halley approaches the sun, these particles burn off creating a magnificent tale that can reach over 5 million miles in length.

Last made its closest approach to the Earth on April 10-11, 1986. Reached its aphelion (or farthest distance from the sun) on December 9, 1923. Halley's Comet is expected to make its next perihelion (closest approach to the sun) on July 28, 2061. Depending on how young you were in 1986 and saw the comet, it is possible to see it a second time in a lifetime.

Halley's Comet travels at speeds as slow as 2000 mph when it is at its farthest distance from the sun. As it gets closer to the sun, its speed increases and can fly past the sun and back into deep space at speeds of over 120,000 mph, reaching a maximum speed of 122,000 miles per hour.

While Halley's Comet was first observed in 240 BC, it was closely observed by spacecraft for the first time in 1986.

Perihelion months of past and future Halley's Comet passing's.

	Year	Month
1.	240 BC.	May
2.	164 BC	November
3.	87 BC	August
4.	12 BC	October
5.	66	January
6.	141	March
7.	218	May
8.	295	April
9.	374	February
10.	451	June
11.	530	September
12.	607	March
13.	684	October
14.	760	May
15.	837	February
16.	912	July
17.	989	September
18.	1066	January
19.	1145	April
20.	1222	September
21.	1301	October
22.	1378	November
23.	1456	June
24.	1531	August
25.	1607	October
26.	1682	September
27.	1758	March
28.	1835	November
29.	1910	April
30.	1986	February
Future Visits		
31.	2061	July
32.	2134	March
33.	2209	February